The Inseparable Bond: Science, Technology, and the Modern Society

Sanobar

Contents

1 Introduction

1.1 Reproducibility in Science

In the modern society of the 21st century, science and technology are becoming increasingly embedded in the everyday life of people all around the world. Driggers (2011) stresses the importance of science and technology by pointing to the many times in history that these components have contributed to moving our society forward and how they will keep on doing so. Similarly, Fenster & Gopp (2020) emphasize the important role science plays, and will continue to play, for the continued development of our society, something they claim has become even more critical and apparent in 2020 as the Covid-19 pandemic is casting its shadow over the continents of the world.

The term *science* is very broad. Obviously, the scientific community is vast and encompasses many different fields that all play important roles in shaping the world. Even though these fields are inherently different on many levels, they all work according to some basic principles that can be applied to all disciplines and branches of the scientific community. One example of these fundamental principles is openness: At its core, science depends upon openness and the *"willingness of individual investigators to accept constructive criticism of work that has been conducted"*, primarily with the purpose to promote the advancement of knowledge (Committee on Science, Engineering, and Public Policy (U.S.). Panel on Scientific Responsibility and the Conduct of Research, 1992, p. 154).

Before a scientific claim can be accepted as a general fact, it must be possible for independent researchers to reproduce the results on which the claim was made to either corroborate or dispute it; after all, *"confidence in results is of paramount importance to the broad scientific community"* (McNutt, 2014, p. 229). Nosek et al. (2015, p. 1422) argue that *"transparency, openness, and reproducibility are readily recognized as vital features of science"*, and that most scientists accept these features as *"disciplinary norms and values."* Freire et al. (2016, pp. 109–110) echo this sentiment and claim that openness, transparency and reproducibility should be integrated in the everyday work of researchers as these practices *"give confidence in the work; help research as a whole to be conducted at a higher standard and be undertaken more efficiently; provide verifiability and falsifiability; and encourage a community of mutual cooperation."* Furthermore, Freire et al. (2016, p. 110) claim that they also lead to a *"valuable form of paper, namely, reports on evaluation and reproduction of prior work. Outcomes that others can build upon and use for their own research, whether a theoretical construct or a reproducible experimental result, form a foundation on which science can progress."*

Based on this symbiotic relationship between openness, transparency and reproducibility, the ability to reproduce any scientific results to validate a claim emerges as another key principle that is commonly seen as fundamental for modern science, regardless of the scientific field that makes up the context in which the claim is being made. As such, any researcher that wants to contribute to the production of knowledge in our society should ideally make sure that all necessary reproducibility requirements are met when reporting the results of a study.

According to a U.S. National Science Foundation (NSF) subcommittee on replicability in science (as cited in Goodman et al. (2016)) reproducibility is a minimum necessary condition for a finding to be believable and informative and refers to the ability of a researcher to duplicate the results of a prior study using the same material as in the original study. Goodman et al. (2016)

notes that the importance of multiple studies corroborating a given result has been acknowledged in basically all areas of science, something that is mirrored by a substantial increase in the number of publications recorded in Scopus[1] that mention, either in the title or abstract, at least one of the following expressions: Research reproducibility, reproducibility of research, reproducibility of results, results reproducibility, reproducibility of study, study reproducibility, reproducible research, reproducible finding, or reproducible result. The quantity of annual publications that meet these conditions has increased from less than 50 in the 1970's to over 300 as of 2014; the field of clinical medicine accounts for the majority of publications that meet the aforementioned requirements. (Goodman et al., 2016) This observation seems to mirror an increased awareness of the importance of reproducibility to corroborate results reported by other scientists.

However, when discussing the term reproducibility in relation to the field of computational science, it is worth noting that it was originally not applied to corroboration, but to transparency. According to Goodman et al. (2016), the term was coined by American computer scientist Jon Claerbout who associated it with a software platform and procedures that allow the reader of a paper to see the complete processing trail, starting from the raw data and code all the way to the final figures and tables as they are presented in scientific publications. Goodman et al. (2016) argue that this kind of reproducibility requires data scientists and researchers to at least share their analytical data sets, relevant metadata, analytical code, and executed software.

1.1.1 Definition

Gundersen & Kjensmo (2018) state that there is a general consensus among researchers in the artificial intelligence field that empirical results should be reproducible, although the meaning of reproducibility is neither clearly defined nor agreed upon. As Goodman et al. (2016, p. 1) puts it: *"The language and conceptual framework of 'research reproducibility' are nonstandard and unsettled across the sciences."* Similarly, Beam et al. (2020) argue that reproducibility is a minimal requirement in the process of creating new knowledge and in the striving for scientific progress, although defining exactly what constitutes reproducibility in a scientific context appears to be a complex issue that has been the subject of considerable effort by both individual researchers and different organizations. When discussing the concept of reproducibility, one must first make sure to properly distinguish between the notions of reproducibility and replicability. According to definitions by Beam et al. (2020), a study is *reproducible* if an independent group can obtain the same results as those observed in the original study, assuming the group has access to the same underlying data and analysis code as used in the original study. A study is *replicable* if an independent group studying the same phenomenon reaches the same conclusion after performing the same set of experiments or analyses on new data.

As such, a clear and consistent definition of reproducibility is necessary for a meaningful discussion and analysis of reproducibility in this paper. Hence, throughout the paper we adhere to the distinctions of reproducibility and replicability as proposed by Beam et al. (2020); the terms as they are used and distinguished in this book is summarized in Table 1. While Beam et al. (2020) are primarily concerned with these concepts within machine learning, the authors' definition of the terms is well in line with what NSF proposes. Our intention with this book is not to determine any universal definition of the term reproducibility, as this undertaking would simply amount to an exercise in semantics. Hence, the meaning of the word presented here might differ from what can be found in other works that deal with these questions. For example, Stodden (2011, p. 22) suggests that reproducibility is related to the process of regenerating the results with *"at least some independence from the code and/or data associated with the original publication."* Drummond (2009) shares the definition by Stodden: He defines replicability as repeating the experiment in the exact same way as in the original study, while reproducibility is the process of recreating the results using another kind of experiment. The definitions of reproducibility proposed by Stodden and Drummond are both different from the one suggested by Beam et al. (2020), and both are more in line with the notion of replicability as defined by the latter. This observation showcases how these terms are not universally defined nor agreed upon in the scientific community, and a shorter overview of the terminology debacle surrounding reproducibility and replicability is presented by Plesser (2018).

Table 1: *Definition of the notion Reproducibility used in this book juxtaposed against the definition of Replicability. Included for the purpose of clarity since inconsistencies often occur when these terms are discussed by researchers. The definitions included here are those suggested by Beam et al. (2020).*

Notion	Definition
Reproducibility	The ability of a researcher to duplicate the results of a prior study using the same data and tools as the original investigator.
Replicability	The ability of a researcher to duplicate the results of a prior study if the same procedures are followed but new data is collected and analysed.

Furthermore, we agree with much of the existing literature that reproducibility is not a boolean variable that represents an "Either/Or" condition under which an experiment is either fully reproducible or not reproducible at all (Isdahl & Gundersen, 2019). We believe that it should be seen as a one-dimensional variable that must be measured on a scale where it can range from multiple *degrees* of reproducibility. As such, we concur with Peng (2011) who suggests that reproducibility can be perceived as a continuous variable on a *spectrum of reproducibility*, on which a specific experiment can range from being not reproducible at all, to the "Gold standard" of reproducibility where linked and executable code and data accompany the research publication that describes the experiment. As explained by Peng (2011, p. 1226), the spectrum of reproducibility is used to explain the gap that arises in the *"scientific evidence-generating process between full replication of a study and no replication."* The spectrum of reproducibility is illustrated in Figure 1 and this key concept is revisited later in this paper.

1.2 Problem Context – A Reproducibility Crisis

Even though reproducibility is such an important requirement to meet in the process of conducting good science, it is not an entirely easy condition to fulfill. This statement is illustrated by

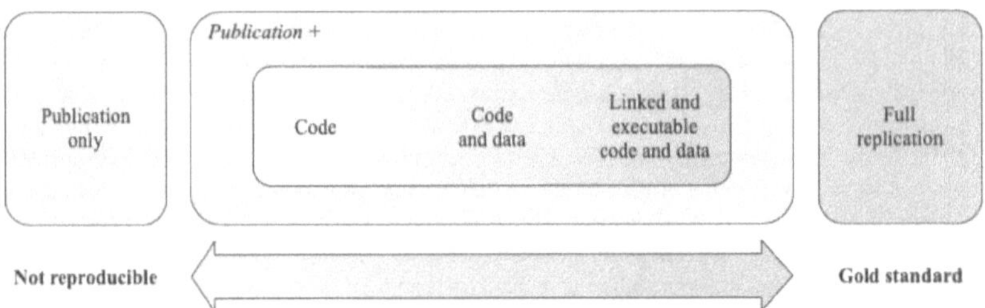

Figure 1: *The spectrum of reproducibility showcasing the relationship between the degree of reproducibility and the availability of the original researcher's code and data. Here, an experiment is not reproducible if no code or data accompanies the publication in which the experiment is described, while the "Gold standard" of full reproducibility is achieved through linked and executable code and data. Illustration adopted from Fig. 1. in Peng (2011).*

numerous articles in recent years that mention a *reproducibility crisis* that is taking place in modern computational sciences. Stodden (2011, p. 21) claims that it is *"impossible to believe most of the computational results presented at conferences and in published papers today. Even mature branches of science, despite all their efforts, suffer severely from the problem of errors in final published conclusions."* While Stodden's statement refers to computational science in general and does not single out a specific field as being particularly exposed to reproducibility issues, a direct crisis has been observed specifically in areas where machine learning (ML) and artificial intelligence (AI) methods are becoming increasingly popular to use for data exploration and analysis. Rupprecht et al. (2020) point out that the exploratory and iterative nature of data science projects leads to a reproducibility crisis in modern data science, while pointing to ML and AI as being especially exposed to this kind of problem. Rupprecht et al. (2020, p. 3354) argue that *"without knowledge of how a result was produced or model was trained, they cannot be trusted. One cannot repudiate, reproduce, reuse, or improve on previous results if their lineage is not understood. This lack of trust hinders widespread adoption, collaboration, and continued refinement."* Obviously, this is problematic considering that ensuring reproducibility of scientific results is a cornerstone of scientific research.

One aspect that is critical for reproducibility in ML and AI (and computational sciences in general) is public access to the underlying analysis code and data that were used by the original researcher to reach the reported conclusions (Brinckman et al., 2019; Chard et al., 2015; Donoho et al., 2009; Eglen et al., 2017; Peng, 2011). How should these key components of a machine learning life cycle be documented in a viable way to promote reproducibility? Brinckman et al. (2019) and Chard et al. (2015) agree that the key to success does not lie in the classical way of documenting one's methods in scientific publications and journals. Brinckman et al. (2019) highlight that scholarly research has evolved significantly in the past decade, while the methods that are typically used to capture and disseminate research processes have not; they claim that the so-calledc scholarly publication is mostly unchanged since the inception of the scientific journal in the 1660's. Brinckman et al. (2019, p. 854) note that this *"disparity"* between modern digital research and the conventional methods used for validation has resulted in many arguments that the scholarly publication is no longer an instrument good enough to *"verify, reproduce and extend scientific results"* (for reference, see Peng, 2011; Kratz & Strasser, 2014; Alsheikh-Ali et al., 2011; Donoho et al., 2009). As today's modern scientific landscape is *"littered with a vast array of powerful cyberinfrastructure for acquiring, storing, analyzing, publishing, and archiving data"* (Brinckman, 2019, pp. 854–855), there must be adequate methods in place for disseminating, validating and verifying computational research that accommodate this reality. According to Brinckman et al. (2019), current approaches do not live up to the documentation and reproducibility requirements set by modern computational sciences. And while there has been an increasing recognition of the need to share all the aspects of the research process (and efforts made to promote data availability and transparency, see for example Chard et al. (2015) and Collins & Tabak (2014)), scholarly publications are still often detached from the code and data that generated the results (Brinckman et al., 2019).

While the availability of data and code is crucial in computational sciences, this is not enough when it comes specifically to the reproducibility of machine learning experiments. For example, Beam et al. (2020, p. 305) suggest that the challenges of reproducing a machine learning model trained by another independent research team can be a challenging endeavour even with *"unfettered access to raw data and code."* There are several explanations to this unfortunate phenomenon, one being that machine learning models have an enormous number of parameters that must be either learned through the use of data or set manually by the researcher. Many times, manual documentation of the configuration that was used when setting up machine learning models is not a feasible option considering that millions of parameters could be involved, while many decisions are also made in a silent fashion through default parameters that are selected by software libraries and not the human user. As such, these default parameters may differ between libraries and may even differ between different versions of the same library; a possible consequence is that researchers who use the same code but different versions of a specific library could reach different conclusions in the end if key parameters are assigned different values. (Beam et al., 2020) Manual documentation of model configuration and involved metadata might be possible in theory, but this is clearly not the case for most real-world cases due to the intricate nature and underlying logic of machine learning models.

Furthermore, Rupprecht et al. (2020) underline the complexity of data science pipelines as they are made up of multiple steps covering many different activities such as data integration, preparation, cleaning and model selection, where each step involves parameters that may be explored and tuned to achieve the desired results. As such, there will be multiple iterations over the same data set and the use of different tools for the pipeline's different steps; tracing these steps is important but is made difficult as data science is often an *"unstructured, ad-hoc process"* (Rupprecht et al., 2020, p. 3355), and is made even more challenging when the number of generated data sets and the complexity of the pipelines increases. For example, these obstacles make it difficult to roll back pipelines to a previous working state when errors are identified in the current state, and this is of high importance in a business context.

In practice, the reproducibility issues of ML and AI are especially clear in certain fields where ML algorithms have great potential, but fail to make it past the drawing board due to lack of transparency, validity and reproducibility. One such an example is the medical field where AI has become a hot topic in respiratory medicine and critical care (The Lancet Respiratory Medicine, 2018). When the UK National Health Service released its initial code of conduct for data-driven health and care technology in 2018, the purpose was to *"encourage the use of tools that improve the quality and safety of care, but also to ensure data-driven technologies are harnessed in a safe, evidence-based, and transparent way"* (The Lancet Respiratory Medicine, 2018, p. 801). Specifically, the tenth principle of the code of conduct is related to reproducibility as it highlights the need for transparency in algorithm learning models to build trust among users. However, this transparency requirement seems to be difficult to fulfill in medical science. Kondrateva et al. (2019) highlight how diagnostics of psychiatric and neurological disorders such as depression and epilepsy can be a problem where specific brain region properties could be extracted and used for classification with machine learning. However, even though a significant amount of studies is being conducted on this topic, the majority of them have small clinical or industrial potential since most studies cannot be reproduced, and therefore not validated and trusted. According to Beam et al. (2020), machine learning methods that are gaining ground in clinical settings present unique challenges and obstacles to reproducibility and that ML models should be reproduced (and ideally replicated) before it can be even considered to be deployed safely and effectively.

Arguably, the medical field appears to be especially exposed to the problems that arise from what many refer to as the reproducibility crisis with regard to machine learning. Nicolas Rougier (as cited in Hutson, 2018, p. 725), a computational neuroscientist at France's National Institute for Research in Computer Science and Automation in Bordeaux, believes that *"people outside the field might assume that because we have code, reproducibility is kind of guaranteed. Far from it."* The statement by Nicolas Rougier is echoed by the findings presented by Gundersen & Kjensmo (2018): They point to research by Braun & Ong (2014) who argue that reproducibility should, in theory, be possible to a higher extent in ML since everything is available on a computer, but apparently the percentage of research that is reproducible is still not higher for ML and AI than for other areas of science. Furthermore, Gundersen & Kjensmo (2018) conducted a survey on 400 AI papers that showed that the documentation of the included research results was not good enough to reproduce reported results and that no paper included in their survey could be considered fully reproducible.

Much of the reproducibility challenges found in machine learning can be traced to the black-box nature of machine learning models, meaning that *"data goes in, decisions come out, but the processes between input and output are opaque"*, something that is especially true for neural networks where data undergo complex transformation in multiple layers of the model that will eventually make the model behave in unpredictable ways (The Lancet Respiratory Medicine, 2018, p. 801). This results in models that are opaque, which is the opposite to the ideal state of transparency. This complex nature renders many sophisticated machine learning models simply too complex for humans to fully understand (Azodi et al. (2020); Barredo-Arrieta et al. (2019)). Due to this intrinsic characteristic of complex machine learning models, they are left unexplainable, not only to those who seek to reuse it for validation purposes, but also to the original creator of the model. When the so called *explainability* of a model is low, it becomes challenging to fundamentally understand why a model takes certain decisions and functions the way it does (Ariza et al., 2020; The Lancet Respiratory Medicine, 2018). That this phenomenon negatively impacts transparency and reproducibility is probably easy to understand, and it is clearly an issue when it comes to real-world use-cases of ML and AI, which has not gone unnoticed in recent years; Miller (2018) claims that the notion of explainable AI has seen a resurgence in recent years as researchers and practitioners are seeking to increase the transparency of their models. There is no doubt that explainability is tightly connected to transparency and, therefore, one could argue that it is also closely related to the notion of reproducibility: A low level of explainability in a certain model could have a negative impact on the opportunities to reuse and validate the model in a fair and appropriate manner, hence damaging the degree of reproducibility of experiments in which the model has been either developed or used.

The importance of transparency and explainability of machine learning models is exemplified well by Barredo-Arrieta et al. (2019). They state that in the *"general context of Artificial Intelligence (AI) development, it is well known that when models are not furnished with self-explanatory characteristics, they may guide to pitfalls. A model that is not understandable can lead to incorrect conclusions, even with correct outcomes"* (Ibid, p. 2232). They go on to illustrate this by pointing to a number of real-world instances reported in media where models possessing a low explainability have resulted in serious consequences for people in their everyday life, with no bad intentions from the model creators themselves (Cush, 2016; Angwin et al., 2016; Parsons, 2016; Rocha, 2016; Reese, 2016). These examples all showcase situations where ML models have behaved in unpredictable ways due to a low explainability and are all related to data modelling aspects such as ethics, accountability and transparency. While this paper has mostly discussed the importance of reproducibility in academia and medical contexts so far, note that all of the aforementioned machine learning incidents occurred outside of the academic sphere, which underlines the importance of validating machine learning models through a process of reproduction before commercial deployment if our society truly wants to reap all the benefits that could come from ML and AI technology. Arguably, these instances of bad model behavior could have been avoided if the models had been reproduced and validated prior to deployment, but this was probably not possible due to an insufficient degree of explainability and transparency.

There should be no question by this point that reproducibility in machine learning contexts is important for society as a whole considering that machine learning systems and algorithms are becoming increasingly popular to deploy in a wide range of situations and applications. Efficient management of the reproducibility crisis in machine learning becomes increasingly important because of this increased interest in ML and AI solutions; to promote reproducibility, initiatives must be taken in serious endeavours to ensure that reproducibility requirements are met so that machine learning models can be validated and deployed safely in settings where they could be of real value.

1.3 Purpose and Aim

The problem discussion presented above highlights the dangers that the reproducibility crisis can be in the context of machine learning and artificial intelligence. Luckily, recommendations and guidelines have been suggested for promoting reproducibility and transparency as an important step in the right direction to lift the scientific community out of the reproducibility crisis (see Section 2). Such recommendations, guidelines and best practices might sound easy in theory, but the next question is how researchers should accommodate these recommendations and guidelines in practice. As mentioned in Section 1.2, the case can be made that documenting the research process in scientific publications is no longer a tool that is powerful enough to ensure reproducibility of computational research according to Brinckman et al. (2019), including machine learning research. Obviously, other methods and tools are necessary to meet the suggested recommendations and guidelines.

Instead, programmatic tracking and version control have been explored as efficient methods for increasing the level of reproducibility of machine learning experiments and pipelines. Rupprecht et al. (2020) suggest incorporating so-called *provenance* into execution environments for data science pipelines, a concept that allows for tracking of all actions and processes taking place throughout the entire machine learning workflow; in other words, a provenance model should be able to record and store the lineage of the pipeline output in its entirety. For example, the lineage of a machine learning model may track variables such as the input data, the algorithms, the number of iterations, and the configuration parameters, which are all components that need to be recorded and stored to ensure future reproducibility and high transparency of the research documentation. These are here on referred to as provenance systems, and such systems should be used to enable widespread adoption, collaboration, reusability and continued refinement of machine learning models. In short, a provenance system can be seen as a tool with the purpose to promote reproducibility by documenting key data in an automatic or semi-automatic manner, saving the data, code and other digital artifacts so that the lineage of processed machine learning pipelines and experiments can be easily revisited by independent users in the future. Various open-source platforms available for public download and use have integrated provenance features as they are described by Rupprecht et al. (2020). Two such examples are MLflow and Pachyderm, both of which are open-source and allow for detailed lineage tracking of machine learning models and related data. These, among others, are presented later on in this paper.

While provenance systems similar to these can manage many issues related to a low degree of reproducibility, there are some aspects related to machine learning workflows and artifacts that such a system would not be able to cover. Examples of this might include user-defined text information such as motivations or descriptions as to why a particular ML model is used for a particular problem, details of the data that was used for model experiments, or general comments regarding ethical considerations that need to be taken into account when using a certain model. As such, documentation of this kind of information can be seen as a way to increase the explainability of a machine learning model. Obviously, recording this kind of information requires the ML practitioner to consider manual documentation, but this is not an entirely simple process: On one hand, as previously mentioned in Section 1.2, manual documentation might not be a viable option when it comes to ML and AI experiments; on the other hand, in cases where manual documentation is possible there might be uncertainties regarding what kind of information that needs to be documented. Even though existing provenance systems might have support for manual documentation through tools such as Jupyter Notebook, it would still lack a standardized and structured framework that indicates what should be documented since notebooks only provide free form text (Isdahl & Gundersen, 2019). Based on this, there is clearly a need to use frameworks for standardized documentation that allow machine learning practitioners to efficiently communicate and share information that can increase the transparency of ML model. Examples of such frameworks are *Datasheets for Datasets* and Google's concept of *Model Cards*, both of which are presented in further detail at a later stage of the paper.

Based on the problem discussion presented in Section 1.2, we want to stress the importance of the following key terms when it comes to machine learning reproducibility: Provenance, Transparency, and Reusability. While these terms are related to each other in the context of machine learning, there are still important differences between them that make it valuable to separate them as three separate conditions for reproducibility. We argue that neither provenance systems nor standardized documentation frameworks in isolation can fully accommodate the three identified requirements for reproducibility. Rather, a combined approach might be necessary to fully reach a high degree of reproducibility when working with machine learning models. This would create a solution that combines the strengths of both tools while minimizing the risk of excluding important information that is necessary for reproducibility and transparency. Therefore, in this paper we present a technical solution that combines the characteristics of a provenance system with an API that allows for a standardized and efficient way to document textual information that the system cannot log automatically.

The solution is implemented in the cloud-native, open-source machine learning platform known as STACKn[23] that runs on Kubernetes. STACKn is used in MLOps contexts for efficient and rapid deployment of collaborative machine learning models, and can be used in any environment that has a Kubernetes cluster running. In its state prior to the writing of this book paper, the STACKn application lacks organically integrated provenance and transparency features that are necessary for a high degree of reproducibility and reusability of machine learning workflows and models. Since many of the stakeholders and potential users of STACKn are individuals who engage in both academic and commercial business settings and activities, many of the them will demand and depend on a high degree of reproducibility and transparency of their ML experiments they conduct in STACKn; they might even request a high degree of reproducibility as a *minimal requirement* in their day-to-day machine learning experiments. Based on this, it is critical to incorporate features in STACKn that organically promote reproducibility, reusability and transparency in a user-friendly way, creating a clear relevance in choosing STACKn as the subject for this book paper. More details on STACKn are included in Section 3.

Using the problem context as described above as a backdrop, the purpose of this book paper is to evaluate the possibility of introducing features for reproducibility support in STACKn, and architect and implement a technical solution in the system that can work together seamlessly with the existing platform architecture. The final solution builds upon accepted recommendations and guidelines for ensuring machine learning reproducibility, reusability and transparency as suggested in literature and by other stakeholders. Our ambition is to introduce reproducibility features and technical characteristics in STACKn that are commonly seen in existing provenance systems and that meet current recommendations and guidelines for promoting reproducibility, as well as a built-in feature that facilitates and standardizes the process of documenting the end-to-end ML workflow when using STACKn for ML m odels. The final aim is to present a solution for reproducibility in STACKn that works well and could actually be of value for the users of STACKn with regards to requirements for reproducibility and transparency. To make the solution valuable for STACKn users, note that we also attempt to make the solution user-friendly, although this is a secondary goal and is not evaluated. As such, the main focal point of this paper is the functionalities and the technical features, and investigating the usability of the product could be an entirely different project in itself. By introducing not only characteristics such as lineage tracking typically seen in provenance systems, but also a built-in documentation API that allows for a streamlined and standardized way of providing transparency to the machine learning workflow, we believe that the concept of the final solution as presented in this paper is unique compared to other open-source tools that exist for the purpose of ML reproducibility. Since we do not know whether both provenance features and documentation frameworks are needed for a high degree of reproducibility, part of the book purpose is also to evaluate if this is the case.

1.4 Research Questions

The purpose statement and aim of this book presented in the previous section lead to the research questions stated below. These serve as a basis and focal point for the remainder of the book and is answered at the end of this paper.

   (i) How can we integrate provenance features typically seen in existing ML platforms, such that ML experiments can be easily reproduced by tracking the lineage of models that are created in STACKn?

  (ii) How can we integrate an API for easy manual documentation in STACKn that encourages users to conduct transparent textual reporting of model characteristics, while increasing reproducibility in the process?

(iii) To what *degree* is it possible to achieve support for reproducible empirical research in STACKn, assuming adherence to the definitions of reproducibility as stated in Section 1.1? More specifically, where on the spectrum of reproducibility as presented in Figure 1 can we place experiments conducted in STACKn using the implemented solutions? Is it necessary to apply both provenance tracking features and manual documentation to obtain a high degree of reproducibility? How does it compare to reproducibility support seen in other similar machine learning platforms?

As we explore these research questions for the remainder of the paper, our intention is to make it as clear as possible as to how we reach our final conclusions. To this end, the paper consists of a number of relevant sections and subsections, each part playing a necessary and critical role in explaining choices made throughout the project and providing details on how we reach our conclusions. As such, the structure of the paper is as follows.

First, Section 2 presents a collection of recommendations and guidelines that have been proposed by various stakeholders to promote reproducibility in computational science. In Section 3 we describe the STACKn platform in more detail, including its architecture and how its components work; it primarily provides a bigger picture of STACKn to give a broader understanding of the application from a higher level. In Section 4 we present a sample of existing open-source ML platforms that have implemented reproducibility features to some extent, primarily with the purpose to showcase the tools that have inspired our own solution in STACKn. Also, a couple of frameworks for standardized model reporting and transparency is described in this section. In Section 5 we give a detailed description of how we implement reproducibility in STACKn through software engineering, and explain the architectural design choices we make. In Section 6 we describe the process for evaluating the reproducibility features we integrate in STACKn, followed by Section 7 in which we analyze the degree of reproducibility we attain in the evaluation phase. Finally, in Section 8 we present the answers to our research questions and our conclusions, and provide a discussion of our contributions, both in terms of mitigating the reproducibility crisis and to the improvement to STACKn. We also present final notes on the weaknesses of our implementation and suggestions on future work.

2 Towards Reproducible Machine Learning

In the discussion in Section 1.2, we presented what many refer to as the reproducibility crisis, an overview of its causes, and some consequences it has brought upon the field of machine learning and artificial intelligence. Furthermore, we showed that the consequences of the reproducibility crisis are tangible in many settings where ML and AI technology is being utilized for various purposes, not only in academia.

Luckily, Brinckman et al. (2019) state that different measures are being taken at all stakeholder levels – including researchers, software developers, funding agencies, and publishers – to increase the transparency and reproducibility of computational science. This section discusses a selection of recommendations and practices that have been suggested by some of these stakeholder as remedies to the reproducibility crisis in the pursuit to ameliorate irreproducibility in computational research.

Since machine learning is a sub-field of computer science, many of the recommendations and best practices for reproducibility included here are not directed specifically towards machine learning. Nevertheless, what is included here is equally important to machine learning as any other sub-field in the computational sciences.

2.1 Recommendations and Guidelines

Multiple recommendations for increasing the degree of reproducibility have been presented in literature in recent years. For instance, Peng (2011) argues that the availability of code and data is a minimum requirement for ensuring the reproducibility of one's research. Peng (2011) claims that the concept of reproducibility falls short of full replication since it calls for the same data to be analyzed again, rather than focusing on the analysis of independently collected data. As such, Peng's (2011) definition of reproducibility is inherently different from replicability; this goes well in line with our distinction between reproducibility and replicability previously presented in Table 1, in which replicability was said to be attained if the same results are possible to attain when new data is collected and analyzed. As mentioned earlier in the paper, Peng (2011) suggests that the degree of reproducibility of an experiment can be ranked on a spectrum of possibilities that indicates an experiment's degree of reproducibility, a concept that was illustrated in Figure 1. Peng (2011, p. 1226)) points to a review on microarray gene expression analyses that found that studies were either *"not reproducible, partially reproducible with some discrepancies, or reproducible"*, and that this range could be explained by the availability of data and metadata.

Building on the spectrum of reproducibility as presented by Peng (2011), Tatman et al. (2018) suggest an updated version of the spectrum that is more in line with current practices for sharing code and data, and that is more suitable for machine learning experiments. They claim that it does not make sense to only distribute code without data in ML settings, which, in contrast, is recommended in Peng's spectrum. Tatman et al. (2018) suggest three levels of reproducibility indicating what should be shared for each level: 1) Low reproducibility: Finished paper only; 2) Medium reproducibility: Paper, code and data; and 3) High reproducibility: Paper, code, data and environment. These levels of reproducibility are, according to Tatman et al. (2018), equivalent to the three levels included in Peng's original spectrum of reproducibility. As such, the principle is still the same as in the original spectrum of reproducibility, just with small modifications to what exactly should be shared in each reproducibility level.

Similarly, Stodden & Miguez (2014) argue that a necessary response to the reproducibility crisis is to publish the underlying data and code that generated the results. Stodden & Miguez (2014) have proposed a list of best practices for practitioners of computational sciences with the purpose to promote reproducibility, facilitate innovation through the reuse of data and code, and to enable broader communication of the output of computational scientific research. A selection of best practice principles for computational science as suggested by Stodden & Miguez (2014) are listed below. Note that the practices that are presented here are only a sample of those proposed by Stodden & Miguez (2014).

- **Open licensing should be used for data and code.** Refers to making the data and code available and open to the public for reuse to the largest extent possible. Stodden & Miguez (2014) recommend using open licensing[4] as one way to make this possible in a legal sense.

- **Workflow tracking should be carried out during the research process.** This refers to provenance and workflow tracking, and also the importance of publishing the environments in which the computational experiment is performed. These tools are important for reproducibility and also minimize the burden on the researcher; Stodden & Miguez (2014) suggest using version control systems such as Git in order to make the code publicly available.

- **Data must be available and accessible.** This principle is three-fold: i) First, a minimum requirement is to provide a version for the datasets that are generated or collected; ii) Secondly, raw data, either from observations or accessed from secondary sources, needs to be available since results should be reproduced from the earliest digital data in the experiment; and iii) Finally, some kind of external and redundant data storage is needed in cases when using large datasets that must be stored in external repositories. These datasets should be versioned.

- **Code and methods must be available and accessible.** Using version control systems with a public facing option, such as GitHub or BitBucket, allows for others to infer exactly what code led to which results, and also make their own adjustments and modifications without disrupting the original code. Another important aspect of this practice is environment version control; that is, providing surrounding and supporting software needed to run the code properly.

Furthermore, Stodden et al. (2016) propose similar recommendations for increasing reproducibility of computational research, some of which are included here. First, they suggest that researchers share data, software, workflows, and details of the computational environment that generated the published findings in open trusted repositories. They underline that the *"minimal components that enable independent regeneration of computational results are the data, the computational steps that produced the findings, and the workflow describing how to generate the results using the data and code, including parameter settings, random number seeds, make files, or function invocation sequences"* (Ibid, p. 1240). Unfortunately, it is often the case that only the clean path to the published results are shown, when multiple paths have been explored prior to reaching the results. This means that there might be pitfalls hidden in the published results. They also propose that persistent links should appear in the published article and include a permanent identifier for data, code, and digital artifacts upon which the results depend. Finally, researchers should properly document digital scholarly artifacts to enable reuse in the future. For example, researchers should include proper levels of documentation of software and data to enable independent reuse by someone skilled in the field.

In addition to recommendations presented in literature, several suggested remedies and guidelines have been released by funding agencies and scientific societies to transition to a scientific state of higher reproducibility and transparency. For instance, the NSF requires that used software and data are disclosed in the research that they fund. In the 2020 NSF Proposal & Award Policies & Procedures Guide (National Science Foundation (NSF), 2020), Chapter XI.D.4 contains the following guidelines (among others):

- Investigators are expected to share with other researchers, at no more than incremental cost and within a reasonable time, the primary data, samples, physical collections and other supporting materials created or gathered in the course of work under NSF grants. Grantees are expected to encourage and facilitate such sharing.

- Investigators and grantees are encouraged to share software and inventions created under the grant or otherwise make them or their products widely available and usable.

Additionally, the Advisory Committee to the Computer and Information Science and Engineering Directorate at NSF released a report in 2016 (Berman et al., 2016) in which recommendations for a national data science research agenda were presented. Two of these recommendations are stated according to the following:

- **Invest in research into data science infrastructure that furthers effective data sharing, data use, and life cycle management.** Develop programs that focus attention on critical problems (privacy, inference, provenance, etc.) that remain obstacles to the use of data at scale. Research outcomes should ultimately be translatable to infrastructure that enable access to data in ways that: (i) preserve privacy and other commitments made when collecting the data; (ii) enable researchers to make "unbiased" inferences, or understand potential biases in the data and other data use challenges; (iii) support reproducibility; (iv) support access, provenance, sustainability, and other life cycle challenges; and (v) support research into new hardware/software infrastructures needed to support Data Science research.

- **Support research into effective reproducibility.** Develop research programs that support computational reproducibility and computationally-enabled discovery, as well as cyberinfrastructure that supports reproducibility. Potential research efforts may focus on, for example, mechanisms to extend validation, verification and uncertainty quantification to include reproducibility; software standards (creation, test, curation, etc.); tools for sharing and verifying queries on confidential data; tools to understand links between decision-oriented models and their training data (e.g., emerging artificial intelligence models); etc.

In the 2017 version of the European Code of Conduct for Research Integrity (All European Academies (ALLEA), 2017), a list of good research practices is described for contexts such as research environments and research procedures. Two of these suggested practices, which mention reproducibility, are stated according to the following:

- Research institutions and organisations support proper infrastructure for the management and protection of data and research materials in all their forms (encompassing qualitative and quantitative data, protocols, processes, other research artefacts and associated metadata) that are necessary for reproducibility, traceability and accountability.

- Researchers report their results in a way that is compatible with the standards of the discipline and, where applicable, can be verified and reproduced.

3 STACKn

This section provides a description of the machine learning platform in which reproducibility features are implemented in this book project, namely STACKn. First, we provide a high-level overview of the platform with background and general information, followed by a brief but more detailed presentation of the architectural components of STACKn that are relevant in the context of this paper.

3.1 Overview

STACKn is an open-source and cloud-native platform for collaborative machine learning developed by Scaleout Systems in Uppsala, Sweden. While the platform is developed in collaboration with various partners such as Safespring and Uppsala University, Scaleout remains the main contributing organization behind the project. Being developed as an open-source product with integration of numerous external open-source libraries, it is also evolved by means of community feedback and adoption. With STACKn, MLOps teams can circumvent many of the challenges that arise in collaborative machine learning projects: The platform aims to optimize and accelerate management of the entire machine learning life cycle seen in Figure 2, and can aid MLOps teams with a multitude of ML related operations, from simple model experiments to model deployment through an endpoint for model serving in production environments. STACKn can be deployed and used in any environment that is compliant with the Kubernetes API, both on local hardware and in the cloud. The possibility to deploy the platform both locally and in the cloud is one of the major benefits with STACKn.

By the time of writing and publishing this paper, STACKn is in closed beta but is currently used in various settings, such as the SciLifeLab Data Centre[5] and within the EU-funded project EOSC-Nordics[6].

3.2 Components and Architecture

To provide basic understanding of the features and functionalities of STACKn, this section provides a detailed description of a selected part of the platform's architecture. The basic comprehension provided by this section is critical for understanding the latter parts of the paper where the implementation stage of the software engineering process is presented. Therefore, all components of STACKn are not discussed here, but rather the components that are of highest relevance for this paper.

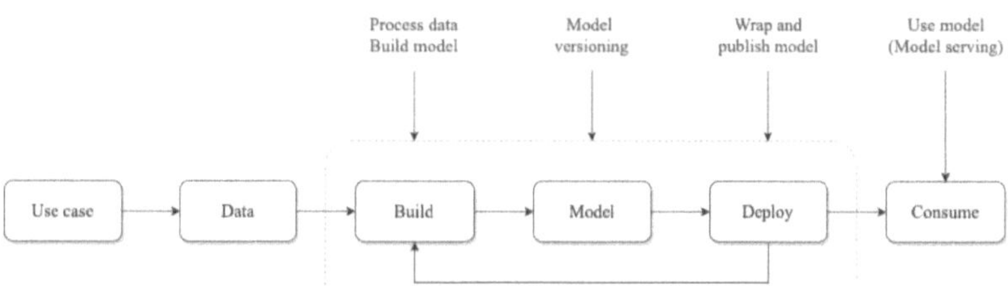

Figure 2: *Illustration of a generic workflow when deploying a machine learning model, from pilot and experimentation to production. The primary purpose of STACKn is to accelerate this process and take machine learning models from proof of concept to production at any scale.*

The components of STACKn that are presented from a high-level perspective here are the CLI, the STACKn Studio App and the Studio API. While a description of how the platform is deployed through Kubernetes is not directly relevant in the context of this paper, it is nevertheless a key concept for STACKn around which its functionalities revolve; as such, it is good to have a basic understanding of how the Kubernetes API is used to manage the application's infrastructure. Therefore, this section begins with a short description of how STACKn is orchestrated through the Kubernetes API. Also note that what is presented here refers to the version of STACKn available on GitHub by the time of writing this paper. Since constant changes are made to the infrastructure and the UI, what is described here might not be true for future versions of STACKn. See Figure 5 for an illustration of a limited section of the STACKn architecture, summarizing what is described in this section.

3.2.1 Kubernetes Orchestration

The STACKn platform is deployed through containerized software units – simply referred to as containers – that packages code and all dependencies so that the platform runs quickly and reliably from any computing environment. A container in this context is comparable to a virtual machine (VM), but much more lightweight. While a VM virtualizes on a hardware level and essentially has a fully fledged operating system running on top of that, containers virtualize on the operating system level. Having multiple VMs running on the same hardware would quickly consume much of the available resources in the system, such as RAM, storage and CPUs. Using containers instead, one can circumvent this issue. (Merkel, 2014) In STACKn, these containers are managed through the use of Kubernetes[7], commonly referred to as K8s: From a high-level perspective, Kubernetes is a Docker container orchestrator that groups the application's software containers into logical units for easy management and discovery.

In STACKn, different application services run in separate Docker containers, which in turn are running in separate *Pods*. A Kubernetes Pod is a group of one or more containers sharing the same resources, as well as a specification on how to run the containers. The major benefit of using the Kubernetes API is that it automatically manages the application services that is running in the

Pods: Some examples of Kubernetes features include automatically restarting containers if they would fail for some reason, and management of load balancing to make sure the deployment is stable in case traffic to a particular container is high. In STACKn, the user interface of STACKn Studio runs in one Pod, the authentication software Keycloak runs in another, and so on. The STACKn component known as the *Chart controller* manages all platform dependent helm charts, which are collection of configuration files that describe the related sets of Kubernetes resources that are initialized within the cluster.

When the STACKn platform is deployed, a Kubernetes cluster is created and a collection of Pods are initialized. When all Pods have been created successfully, STACKn and its components are ready to be used for machine learning projects. See Figure 3 for a list of the initial Kubernetes Pods that are initialized when deploying STACKn. Note that the figure only shows the *initial* set of Pods that are created and that more are added as the user operates in STACKn.

```
NAME                                              READY   STATUS    RESTARTS   AGE
stackn-celery-worker-fc7c656c6-2q9p6              1/1     Running   0          67s
stackn-celery-worker-fc7c656c6-mzg65              1/1     Running   0          68s
stackn-chart-controller-cfd5bfbf6-74m6d           1/1     Running   0          68s
stackn-docker-registry-7b5b56687d-2mbpj           1/1     Running   0          67s
stackn-grafana-85d5dc6c66-dwms8                   1/1     Running   0          68s
stackn-ingress-b4c55c5f5-kjzbx                    1/1     Running   0          67s
stackn-keycloak-0                                 1/1     Running   0          68s
stackn-kube-state-metrics-68577659b-h6cjs         1/1     Running   0          68s
stackn-loki-0                                     1/1     Running   0          68s
stackn-postgresql-0                               1/1     Running   0          68s
stackn-prometheus-alertmanager-798c79bb55-6fxcc   2/2     Running   0          68s
stackn-prometheus-node-exporter-qjbzq             1/1     Running   0          68s
stackn-prometheus-pushgateway-69865c779b-2xnxf    1/1     Running   0          68s
stackn-prometheus-server-66848f97f7-sng47         2/2     Running   0          68s
stackn-promtail-nhvnz                             1/1     Running   0          68s
stackn-rabbit-8478cf9bb7-rbf45                    1/1     Running   0          68s
stackn-redis-774b797f76-87t9d                     1/1     Running   0          67s
stackn-studio-777b4f4665-hxzkh                    1/1     Running   0          68s
stackn-studio-db-7d6c44b94-z4zgd                  1/1     Running   0          68s
```

Figure 3: *Initial resources that constitute the default Kubernetes cluster on which STACKn is running. These are the initial Pods that are generated when the user deploys an instance of STACKn. The "READY" column indicates if the containers are ready and the "STATUS" column indicates in which state the pods are in; in the state of the cluster that is shown here, all containers are ready and all pods are running. As can be seen in the "AGE" column, all pods and containers have been successfully initialized after approximately one minute.*

3.2.2 STACKn CLI

CLI is short for *Command Line Interface* and processes commands to the computer. The *STACKn CLI* is the primary tool for users to interact with STACKn through a command prompt. A variety of different commands is available to the user, facilitating for authentication towards a running instance of STACKn, initialization of a project directory, setting and fetching the active project, creating and deploying models, and so on. The CLI commands assume that the user already has a deployed instance of STACKn in Kubernetes and that the user has created a user account in STACKn. See Table 2 for a list of examples of the commands that are available to use when working with STACKn projects and models.

Table 2: *Examples of CLI commands that are used to work with machine learning models in STACKn. In the table, each command is accompanied with a short description of what it is used for. Note that this is just a small sample of all the CLI commands that are available to use with the STACKn API, and that some of the commands take additional parameters in order to work that are not shown here.*

Command	Description
stackn setup	Command used for logging into the STACKn service deployed by the user. Assumes the user already has a deployment in STACKn and that a user account has been created in this particular deployment.
stackn init	Command used to initialize a generic project structure on the machine where the user operates. This command creates a directory in which users collect all their code and data for experimenting with machine learning models. This should be used at the beginning of a project.
stackn set project	Command used to set a STACKn project as active in the configuration file. This is a necessary command to run before experimenting with models.
stackn status	Command used for retrieving information about which STACKn deployment that is active and which STACKn project that is set as active.
stackn create model	Command used to create a model in the STACKn database. The created model is visible in STACKn Studio inside the active project. When running this command, the user states whether the model should be released as a minor or major version.
stackn create deployment	Command used for deploying a model. This creates Kubernetes resources for the particular model and an endpoint for model serving which the user can call to make predictions with the model.

3.2.3   STACKn Studio

The Studio App service in STACKn is a Web API and consists of two major parts, here referred to as the *Studio UI* and the *Studio Component*. The Studio UI refers to the graphical user interface of the STACKn platform and is built on the Django REST framework[8]. The Studio Component handles the software logic behind the curtains of the UI, incorporating HTML, JavaScript and Python to create both a proper front-end and infrastructure for managing queries to the back-end. Since the service is run on a, often remote, Kubernetes cluster, STACKn Studio can be reached from anywhere; as seen in Figure 3, the initial setup of the Kubernetes cluster contains a pod dedicated to managing the Studio services (see pod labelled "stackn-studio-777b479...").

In the Studio UI, which is available from a browser, the user can create Projects for managing and collecting their machine learning queries and models. To give an idea of how the STACKn UI looks, an example view can be seen in Figure 4: In this figure, the user has created a project titled "MNIST" and entered the project's home page in the browser. The user can navigate the project page to engage in numerous activities through the left navigation bar, some examples include uploading datasets to the S3-compatible object storage MinIO under the tab "Datasets", managing project specific models under the tab "Models", creating custom Jupyter lab sessions under the tab "Labs", and much more.

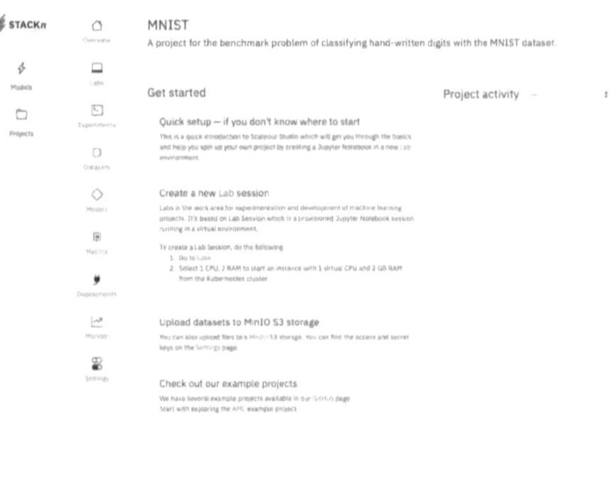

Figure 4: *Home page of a STACKn project titled "MNIST".*

3.2.4   Studio API

When any CLI command of STACKn is called by a user, the *Studio API* (also referred to as the Studio Client) handles the logic behind the commands. This module manages communication between the client side and the server side of the platform, handling requests from the user when operating the CLI. For example, when the user runs `stackn create model`, the Studio API is the first in line to manage the underlying logic by sending a POST request to the Studio App through an endpoint that points to the model database, thus creating a new model instance in the database and publishing it in STACKn Studio. In short, the Studio API tells the Studio App what data that should be retrieved or sent to the database, while the Studio App makes sure that databases are properly updated and that data is visualized as intended in the Studio UI. As such, the Studio API acts as an intermediary between the CLI and the Studio App.

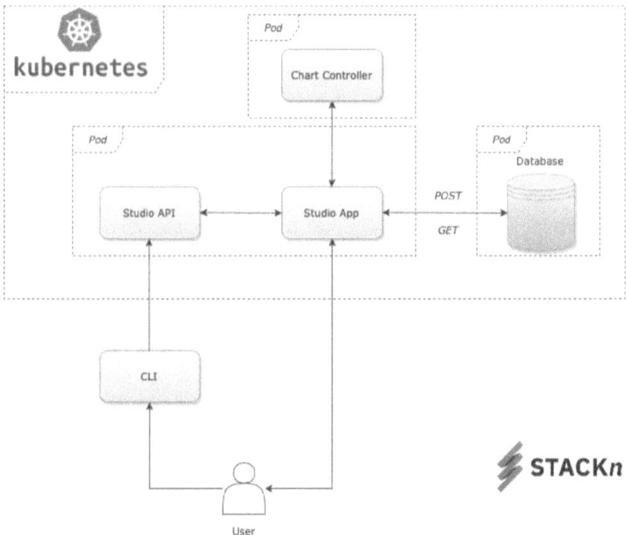

Figure 5: *A simple illustration of an isolated part of the STACKn architecture. This limited section of the architecture illustrates how the Studio API and Studio App reside in the Kubernetes cluster packaged as containers in separate pods, together with the Chart controller and the database. These components communicate with each other, the user and the CLI according to the arrows. The Chart controller is needed to start the Pods when deploying STACKn. Note that the cluster consists of additional pods that are not shown here.*

3.3 Usage

This sub-section presents a simple example of how users can create and deploy a simple model in STACKn, assuming the user has already created a project in Studio. We use the previous example project titled "MNIST" for this purpose when applicable.

First, the user can choose to run experiment code either locally or in a Jupyter lab session that can be created in Studio. A Jupyter lab is the user's main experimentation hub when working with machine learning models in STACKn, providing notebooks for documentation and access to external computation resources. If the user creates a new Jupyter lab, new Kubernetes resources are generated in the cluster. Then, as an initial step before any models can be created, the user must first enter the CLI command `stackn init` to initialize a default project structure in a directory on the local machine. This directory inhabits a number of sub-folders and files that all contain important data and code necessary to perform experiments on a model. For example, one of these files is `train.py`, which the user should populate with the code needed to train the model. Also, a `requirements.txt` file is included in the project structure directory in which the user enters all the software packages that is needed for conducting the model experiments. After this, the

19

user enters `stackn set project -p MNIST` to set the existing project "MNIST" as active in the application's configuration file.

The next step is to download datasets for generating a model and store them in an appropriate folder in the previously initialized project structure. This is followed by any eventual data pre-processing steps, performed through code that should also be stored in files in the project structure directory. The data in its final form is also stored in the project directory. After this, code for training the model is defined and executed, after which the generated model object and weights are saved.

To use this trained model, the methods `model_load` and `model_predict` are implemented in the file `predict.py` to enable model predictions, either locally or by calling a model endpoint. To set up a model endpoint, the user first runs the CLI command `stackn create model -n mnist -r minor` to create a model of a minor version release titled "mnist". The newly created model is published in STACKn Studio. Thereafter, the model is deployed in a default-python environment through the command `stackn create deployment -m mnist -d default-python`. This creates a new Kubernetes Pod for the model and a serving endpoint for the model that is visible in Studio. Using the endpoint URL, a prediction can be made with the model at any time in the future, as long as the endpoint is still running properly in the Kubernetes cluster. By deploying models in this manner, teams can take control of their developed models and use many models in parallel.

4 Tools for Reproducibility and Transparency

This section provides a presentation and description of existing tools that integrate and/or promote features for reproducibility and transparency when used in a machine learning context. While Section 2 discussed theoretical remedies for the reproducibility crisis through recommendations and guidelines for computational researchers, this section describes a selection of tools that have been developed to facilitate reproducibility and transparency in practice. The term *tools* in this context refers to two different kinds of tools that can be used for increasing the degree of reproducibility and transparency in machine learning settings: 1) Open-source ML platforms with integrated provenance system features for lineage tracking and logging, and 2) Frameworks for standardized documentation promoting data and model transparency. The main purpose of this section is to provide a basic understanding of the tools that have inspired our reproducibility solutions in STACKn.

4.1 ML Platforms

The following sub-section is a presentation of existing open-source ML platforms that are similar to STACKn in the sense that they support and simplify development and deployment of machine learning models by offering a wide range of functionalities to manage the machine learning life cycle. However, the chosen platforms differ from STACKn in that they have active support for reproducibility through various kinds of provenance tracking features, which is why they are chosen as inspiration for reproducibility features in STACKn. This is mainly done to give us an idea of how reproducibility features have been implemented in similar systems and what kinds of features that are actually possible to implement. This section does not provide a detailed dissection of each platform in terms of "under the hood" functionalities, but mainly presents an overview of the platforms and the concepts that they are built on. Furthermore, as these platforms are more than just provenance systems and provide features that are not directly related to reproducibility and transparency, only the parts of the platforms that are relevant from the point of view of this paper are described here (although other components of the platforms might be mentioned as well). In other words, the relevant parts to include in this section is how the different platforms have integrated reproducibility features.

The platforms that are included here are *MLflow*, *Kubeflow*, *Pachyderm*, and *TensorFlow Extended*. These platforms have inspired different parts of the solution implemented in STACKn; which part each platform has inspired is clarified in the presentation below. When selecting other platforms to investigate, we based our choices on the following criteria: 1) Clear and sufficiently extensive documentation, 2) open-source availability, and 3) sufficient community adoption. The platforms included here are far from the only ones with integrated reproducibility features that we could have investigated, including Floydhub, Google Cloud ML, Amazon Sagemaker and Azure ML. However, these do not fulfill the aforementioned criteria.

Obviously, there is a possibility that other platforms might have been better choices for our purposes compared to the ones we actually chose (for example due to more similarities to STACKn). However, it would not have made sense to spend too much time on investigating other platforms as this would have been out of scope for this book.

4.1.1 MLflow

MLflow[9] is an open-source platform for the entire machine learning cycle, which includes experimentation, reproducibility, deployment, and a central model registry. MLflow was introduced by Zaharia et al. (2018) as a response to demand from Databricks customers who expressed concerns that ML life cycle problems are often a major bottleneck when launching a model in production. Zaharia et al. (2018) argue that much of the ML carried out in commercial settings today lacks the rigid and reliable engineering processes found in, for example, software development. ML projects rarely come with a clear and concrete end-goal regarding functionality, as is the case in software development. Instead, these kinds of projects rely on experimentation and incrementally increasing model performance. Specifically, MLflow addresses the ML challenges that Zaharia et al. (2018) found to be the most common among ML users, which are described below:

(i) **Multitude of tools:** The amount of software available to developers, be it in ML as well as traditional software engineering, is vast. Hundreds of libraries and tools are available for each step of development. In traditional software development, it is common to pick a single tool for each phase. However, in ML engineering, developers probably want to try several, if not every, available and relevant tool (for instance, pre-processing and model selection) in a single phase of development to maximize performance.

(ii) **Experiment tracking:** The performance of ML models depends on the tuning of a large number of parameters. Keeping track of these can be difficult, no matter if it is a single person or a team working on a model. Hence, tools to track these parameters are extremely useful.

(iii) **Reproducibility:** Since model performance and behavior depend heavily on the input data and training process, reproducibility is absolutely "paramount" throughout ML development. However, without detailed tracking of ML experiments, data scientists face difficulties in getting the same code to work again.

(iv) **Production deployment:** Deploying a model from a development environment to production is challenging for inference and training. For example, there are many different environments that can be used for inference, but no standard way of moving models from any arbitrary library to these diverse environments.

To manage the challenges described above, MLflow offers four components that are vital to the aforementioned ML life cycle:

(i) **MLflow Tracking:** This component is an API and UI used for logging parameters, code versions, metrics, and output files when running machine learning code and for later visualizing the results.

(ii) **MLflow Projects:** Format for packaging code into reusable projects. Included in a project is its environment (including necessary software libraries), the code which constitutes the machine learning program to run, and parameters that can be used when calling the project programmatically.

(iii) **MLflow Models:** Format for packaging machine learning models, both the required code and data, that works with a multitude of deployment tools (for example serving with either batch and real-time inference).

(iv) **Model Registry:** A centralized model store, set of APIs, and UI, used to manage the full life cycle of a machine learning model collaboratively. It provides model lineage, model versioning, stage transitions, and annotations.

The most interesting aspect of MLflow, in the context of this book, is the MLflow Tracking component that can be used to log and query experiment runs; these runs consist of parameters, code versions, metrics and output files known as artifacts. MLflow Tracking comes with a number of logging functions that can be called to store interesting metadata – for example accuracy and regularization value – with the purpose to record the lineage of a machine learning workflow. The Tracking API stores data on the local system by default, but it is also possible to log the data over the network to a server. Users can then query the saved data in an API or through a web UI.

4.1.2 Kubeflow

Similarly to STACKn, Kubeflow[10] is a scalable, portable and distributed Kubernetes-native platform for end-to-end ML workflows that can be used anywhere a Kubernetes cluster is already running. Kubeflow was originally developed by Google and is based on their internal machine learning pipelines, and is labelled as the "machine learning toolkit for Kubernetes". It can be used for multiple purposes, such as building, training and deploying ML models, tracking and comparing ML experiments, automated hyperparameter tuning, and sharing of models between teams, clusters and clouds.

Kubeflow consists of a number of components, including the *Central Dashboard* – the central UI in Kubeflow – and the *Metadata component* that helps the user track and manage metadata that are produced when running machine learning workflows. Metadata refers to information about runs, models, datasets, evaluation metrics and other artifacts. All the metadata that is tracked when running a workflow can be seen in the Artifact store in the Central Dashboard, allowing users to easily compare and revisit ML experiments at any point in time for easy model selection, validation and re-usability. Under the Metadata description in the Kubeflow documentation, three different artifacts and corresponding metadata are shown: Model, Metrics and Dataset. For a Model artifact, Kubeflow can save information such as *Model version*, *Model description*, *Model type*, *Run ID*, *Creation time*, *Hyperparameters*, *Training data*, and *Training framework*. For a Metrics artifact, Kubeflow can save information such as *Version*, *Run ID* and *URI* that points to a file location; the same goes for Data artifacts.

4.1.3 Pachyderm

Pachyderm[11] is a platform for conducting data science experiments reliably and in a reproducible way, creating a good foundation for ML and AI that is "explainable, repeatable and scalable." The idea of Pachyderm is to combine version control for data with tools to build scalable end-to-end ML/AI pipelines through their two main concepts – the Pachyderm File System (PFS) and the Pachyderm Pipeline System (PPS). On top of this, users can choose freely what language, framework and tools they want to use to develop their programs; just like STACKn, the Pachyderm platform is machine learning framework agnostic. Investigations into Pachyderm was mainly conducted with the purpose to get an idea of how data version control can be implemented in STACKn, which is why only the PFS concept mentioned above is explained briefly here, and not the PPS.

Pachyderm's versioned data concepts are similar to the Git version-control system, but with a few exceptions. Pachyderm implements "rich version-control and history semantics" for data, and similar to Git, Pachyderm uses repositories where data is stored and that tracks all changes to the data by creating a history of data modifications that can be accessed later. Usually, each dataset is its own Pachyderm repository which can store all types of data files, including binary files. However, the history of the data is not stored in a file similar to .git files; instead, Pachyderm stores the history of all commits in a centralized location, making it impossible for the user to run into merge conflicts as one often does with Git. Furthermore, a commit in Pachyderm is a snapshot of the data (repository) at any given point in time, and just like in Git every commit gets a unique ID that makes it easy to track which version of the data that the commit refers to.

4.1.4 TensorFlow Extended

TensorFlow Extended (TFX)[12] is an end-to-end platform for deploying and managing ML production pipelines. TFX pipelines consist of several components that are built using TensorFlow's own libraries, one of which is the ML Metadata (MLMD) library that stores and retrieves metadata associated with ML developer workflows. The MLMD library can be used to analyze important parts of the ML workflow; for example, MLMD can help developers check which dataset the model was trained on, what hyperparameters that were used to train the model, which specific training run resulted in a specific model, which version of TensorFlow created the model, and so on.

The MLMD library can save important metadata in the so called "Metadata Store"; for example, the kind of metadata that it stores can be metadata about the artifacts generated through the components/steps of the user's ML pipelines. The Metadata Store provides APIs in order to record and retrieve metadata, both to and from the storage back-end. The main inspiration we take from TFX and its MLMD library is the concept of a metadata store and how it can be used.

The functionalities of the MLMD library and some of its major benefits include the following examples:

- List all Artifacts of a specific type.

- Load two Artifacts of the same type for comparison.

- Recurse back through all events to see how an artifact was created.

- Record and query context of workflow runs.

4.2 Documentation Frameworks

The following sub-section discusses two frameworks for standardized documentation proposed for increasing transparency with regards to the reporting of machine learning data and models. These frameworks are *Datasheets for Datasets* and Google's *Model Cards*. While the ML platforms described above can be used for managing the provenance of production pipelines and tracking the lineage of the ML workflow, they do little in the way of providing a standardized approach for documenting information that the provenance system components cannot manage. As such, there is a clear distinction between the platforms presented above and the two textual frameworks that are described below; clearly, both kinds of tools contribute to a higher transparency in the reporting of machine learning life cycles and workflows, but in different ways.

4.2.1 Datasheets for Datasets

There is currently no widespread standard procedure of documenting the datasets used for model training in the ML community. Since the characteristics of a dataset greatly affect the performance and behaviour of the ML model which it has trained, unclear communication about datasets can have grave consequences if the subsequent ML models are used in high-stakes, commercial domains, such as finance, criminal justice, critical infrastructure. For example, bias in a specific dataset can result in discriminating behavior in the models that are trained with the dataset. These are claims of Gebru et al. (2018) who have proposed "Datasheets for Datasets" to address these aforementioned issues. The datasheet is a sort of questionnaire for dataset creators or users to fill out about the provenance, creation, and usage of a particular dataset. The purpose of these datasheets is to provide transparency to the dataset that someone wants to use to train or validate machine learning models, and by doing so reduce the risk of unintentional misuse of datasets. The datasheets should describe the creation, strengths, and limitations of a dataset, as these characteristics are important to communicate to those who might use the dataset in order to avoid misuse.

Gebru et al. (2018) developed a number of questions that they considered important to be included in a datasheet and that should be answered by the dataset creators if possible. They divided these questions into seven categories: 1) Motivation for dataset creation; 2) Dataset composition; 3) Data collection process; 4) Data pre-processing; 5) Dataset distribution; 6) Dataset maintenance; and 7) Legal and ethical considerations. Gebru et al. (2018) note that all questions are not applicable to all datasets. Some examples of their proposed questions include:

- Why was the dataset created?
- What pre-processing/cleaning was done?
- Are there recommended data splits or evaluation measures?
- If it relates to people, could this dataset expose people to harm or legal action?

The development of the questions was guided mainly by a number of important objectives. These were: 1) A practitioner should be able to decide, based on reading a datasheet, how appropriate the corresponding dataset is for a task, what its strengths and limitations are, and how it fits into the broader ecosystem; 2) The creators of a dataset should be able to use the questions to inspire thought about aspects of dataset creation that may not have otherwise occurred to them.

A datasheet like this is filled out completely by hand and relies solely on human input. In other words, there are no Python packages or modules to aid in the creation of a datasheet, and the only practical way to programmatically facilitate the writing of a datasheet is to provide forms for the dataset creator or user to fill out.

4.2.2 Model Cards

In an effort to further provide standardized documentation practices for communicating the performance of ML and AI models, Mitchell et al. (2019) introduced the idea of so called *Model Cards*, which partly builds upon the concept of datasheets suggested by Gebru et al. (2018) as described above. Mitchell et al. (2019) point to research on how systematic ML biases become very problematic when embedded into machine learning contexts that seriously affect the daily lives of people, such as health care, employment, education, and law enforcement. These biases only become apparent after these systems are put into production when users have a chance to report negative experiences. Although these problems have been known for some time, documentation on how trained ML models perform in different contexts, with intended use cases as well as pitfalls, have been scarce. To counteract this, the Model Card concept has been introduced for ML models; a short, standardized document showcasing a specific ML model's performance, not only in the traditional sense of pure performance metrics, but also concerning ethics, inclusiveness, and fairness. When addressing the problem of ML bias, Mitchell et al. (2019) highlight the importance of making the performance analysis intersectional; that is, analysing the performance across two or more sub-groups of the ML input. A facial recognition model, for example, is to be examined for instance on its images depicting black women, thereby including both people of color, as well as women in the performance analysis. Another example could be images of men in poor lighting conditions, and so on.

Mitchell et al. (2019) suggest nine core areas which a model card should cover:

(i) **Model Details:** Basic information about the model such as who is developing the model, date and version, model type, training algorithms and related parameters, paper or similar resource for more information on the model, citation details, license, as well as where to turn with questions or comments regarding the model.

(ii) **Intended Use:** This section should present what the model was built to do. I.e which use cases and which users was envisioned during development. Here should also be included out of scope use cases, where the particular input data might be confusing to the model. If there are any recommendations for other models more suitable for such input, they should be presented in this section.

(iii) **Factors:** Here, factors that might affect model performance is presented. These factors are split into *relevant* and *evaluation* factors, where the former could cover; a) *Groups*, which refers to distinct categories of the input data, sharing one or multiple similar traits. b) *Instrumentation*, the tools used to capture and collect input data. And c) *Environment*, which states how model performance is affected by the setting in which the model is deployed, for example, how poor lighting conditions can affect the performance of a face-detection model. On the actual model card, relevant factors are presented, along with how they were determined. Evaluation factors describe which and why certain factors are reported, and why relevant and evaluation factors might differ. A for the model relevant factor might not, for example, be available as an annotation on the datasets used to train and test the model.

(iv) **Metrics:** Different metrics should be presented depending on what type of model has been used. Performance metrics could be class labels for some models and a score for others. Along with the performance metrics, a section on what metrics are presented and why these were chosen, should be included, as well as how they were calculated.

(v) **Evaluation Data:** Evaluation datasets should be publicly available for third-party use (previously existing and newly created alike), and the references provided in this section should ideally point to documentation on the composition and source of the dataset alongside this, details on why the specific datasets were chosen and information about any pre-processing can be included here.

(vi) **Training Data:** Since some datasets used for training might be proprietary or protected under a non-disclosure agreement, the level of detail of information provided in the earlier section about evaluation data might be hard to achieve. Nevertheless, as much detail as possible about the training data should be presented here.

(vii) **Quantitative Analyses:** The analysis presented should be broken down, or "disaggregated", by the chosen factors, both in a unitary and intersectional fashion. Unitary in this context refers to how the model performs with respect to each chosen factor, while intersectional refers to how the model performs with respect to the intersection of the evaluated factors.

(viii) **Ethical Considerations:** This section of the model card should describe any considerations that need be taken with regards to ethical issues, and suggestions for solutions to these issues. This could include comments on whether the model has used sensitive data, if the model should be used in risk-critical fields, or how the model could be harmful in any way.

(ix) **Caveats and Recommendations:** The final section of the model card should be related to any identified caveats and recommendations that have not already been covered in the previous sections. For example, one might describe here if the results suggested further testing or whether there are any ideal datasets that could be used for model validation.

Consequently and similar to datasheets, producing a model card cannot be fully automated, and its creation relies on human input. The idea is primarily to provide a standardized way of managing the documentation of important information so as to make it easy for machine learning practitioners to include aspects that are necessary for a high degree of transparency. Therefore, Mitchell et al. (2019) points out that the usefulness of a model card in many cases depends on the integrity of the researcher, meaning that a lot of responsibility lies in the hands of the individual researcher when it comes to transparent reporting. Furthermore, they are aware that the model card concept does not solve all transparency issues, but is simply one of many ways that ML practitioners can use to increase the transparency of the model reporting. They suggest using model cards as a complement to other transparency tools to strengthen the documentation.

However, Fang & Miao (2020) introduced the Model Cards Toolkit[13] recently which provides all machine learning practitioners with a streamlined approach to generate customizable model cards for transparent model reporting. This toolkit can be utilized by anyone using TensorFlow Extended and its MLMD library (see Section 4.1.4).

5 Software Engineering Process

This chapter provides a detailed description of the project's software engineering and development process. As this book paper is advocating for reproducibility and transparency in science (particularly in machine learning), one could argue that the credibility of this paper would be undermined and questioned if no efforts are made to increase the reproducibility of the method utilized for this specific software p roject. Therefore, the intention of this chapter is to show, in as much detail as possible, the steps that are carried out to implement reproducibility features in a machine learning framework. Of course, we cannot document everything in minute detail, but we make an attempt to at least provide an overview of how the reproducibility concepts are implemented in STACKn.

Considering that most machine learning and data science platforms differ in many ways – for example through fundamental differences in the architecture and what libraries that make up the foundation of the systems – we are aware that the steps described in this chapter are not a universal solution to reproducibility issues in ML platforms. But just like we are inspired by other existing platforms to implement the functionalities in STACKn, we hope that our method could at least work as an inspiration to others who wish to integrate similar reproducibility features in their own systems. As such, we believe that a thorough breakdown of the technical components created and the methods used to implement them is a necessity.

A final note on this chapter is to keep in mind that what is implemented in STACKn as part of this project is indeed not a final solution, but first and foremost a prototype that lays the foundation for further development in the future. What is important is that this prototype can showcase how reproducibility can be implemented in a platform for machine learning life cycles. Parts of the code written for this project can be found in the Develop branch[14] in the STACKn repository on GitHub (by the time of publishing this paper, all parts of the code have not been merged with the Develop branch).

5.1 Preparations and Planning

Initially, the focus of our work was directed on gaining an understanding of the different software systems that STACKn is built upon, such as Docker and Kubernetes, while also learning to navigate STACKn's existing code base and structure to see how different files and functions are connected and how they work together. This created a good sense of how new components and features could be implemented in an efficient and way and how they should to interact with the existing software. Arguably, understanding STACKn and its architecture was the most crucial part in preparatory stages of this project. As the intention was to develop and make changes to STACKn locally, we used Telepresence[15] to replace the studio pod in the Kubernetes cluster with a proxy that points to a local container. Telepresence is a tool for debugging one's Kubernetes service locally, making it possible to see local changes immediately in the deployment of STACKn, despite the fact that it was a remote deployment. Understandably, this was also a crucial tool to use to successfully contribute to the development of STACKn.

When working with the software engineering process, we have used an agile approach and have loosely followed the design process described by Sommerville (2016). As such, we began by defining our requirements – what kinds of features need to be implemented in order to introduce support for reproducibility in STACKn? This was followed by an architectural design step, after which we began implementing our features in the existing infrastructure of STACKn. Throughout the development, we also conducted careful testing of the implemented features. However, we will not spend too much focus here on describing our design and implementation process from Sommerville's theoretical perspectives; while we conduct an exhaustive software engineering process, the act of conducting it in an optimal way according to theoretical frameworks is not the main focal point of the paper in relation to the purpose and research questions.

To organize tasks and planning for the software development process, Jira Software[16] is used as a structured way to manage the project. Jira Software is a platform for planning and organizing tasks related to the project and allows for efficient project management using an agile approach, and this tool is an absolutely necessary addition to the workflow of the project considering the project's time-frame and scope. To manage communication efficiently during the development stages, we use various channels on Slack[17].

5.2 Requirements Engineering

Based on standard conventions for reproducibility found in the existing ML platforms described in Section 4 and on best-practices suggested by academic literature presented in Section 2, some features and properties are deemed required and absolutely necessary to implement with the purpose to make ML experiments reproducible in STACKn. A summary of these features can be seen in Table 3 together with a motivation of why each one of them was considered necessary. The motivation for each feature is included to show that the implementation of reproducibility in STACKn has not been done arbitrarily in a vacuum; instead, the inclusion of each feature is motivated by academic research and similar features in other frameworks and ML platforms.

Based on Table 3, we employed the concepts of *User requirements* and *System requirements*

Table 3: *Features that are required or necessary to implement to attain reproducibility in STACKn. Each feature is presented with a short motivation of why it was considered necessary to integrate to promote reproducibility in STACKn. Some of the features are tightly connected, for example "Code availability" and "Provenance".*

Feature	Motivation
Code availability	As highlighted by Brinckman et al. (2019), Peng (2011), Stodden & Miguez (2014), and Tatman et al. (2018), to name a few, making the exact computer code used to conduct an experiment available is a key step for promoting reproducibility in computational science and machine learning. The ability to reuse and inspect code that produced a model is essential.
Data availability	By controlling the versions of the datasets with which ML models are trained, validated, and tested, changes and updates in that data can be tracked through regular code versioning services such as GitHub. The availability of analysis data is, similar to experiment code, a minimum requirement for promoting reproducibility in computational science and machine learning (Peng, 2011; Stodden & Miguez, 2014). The ability to reuse and inspect data, exactly as it was constructed in the original experiment, is essential. Tatman et al. (2018) also emphasize the importance of sharing data *together* with the code.
Metadata storage	Important for the features discussed above, tracking and storage of model parameters, hyperparameters and other metrics generated during experiments have to be documented. This also includes other metadata, such as data about datasets and code. The importance of digital artifacts and metadata is highlighted by ALLEA (2017) and emphasized by existing ML platforms, such as MLflow and the MLMD library in TFX.
Environment sharing	As suggested by Stodden et al. (2016), Stodden & Miguez (2018) and Tatman et al. (2018), the environment in which the experiment was executed should be shared as this can easily impact the exact results of an experiment.
Provenance	To allow independent researchers to revisit and reproduce previously conducted experiments, we must enable long-term storage of experiment related artifacts such as code, datasets and metadata. Such information must, at any point in time, be readily available to anyone who wants to investigate the entire path that have resulted in a specific model. The option to revert to previous versions of a model by tracking its entire origin – that is, to track the *provenance* of the model – is a necessity if ML experiments are to be reproducible. The notion of provenance is used by Rupprecht et al. (2020) and Berman et al. (2016), while Stodden et al. (2016) also discusses the process of tracking the complete path for reaching the final results/models.
Datasheet	Gebru et al. (2018) explains the importance of documenting the experiment data on which an ML model is generated. Mainly to increase transparency and to avoid unintentional biases, which can aid with increasing the reproducibility by communicating to others how datasets can be recreated, how data is transformed and reshaped for training, and so on. This is a part the documentation frameworks that are integrated in STACKn.
Model cards	Mitchell et. al (2019) continues the work on data transparency by Gebru et al. (2018) and proposes Model Cards as an extension to the concept of Datasheets. Model Cards can be used as a way to communicate important characteristics and performance of models to increase the explainability of them, for example to avoid models from being used for inappropriate tasks and communicating the best ways for which the model can be evaluated (which metrics to investigate and so on). Like the Datasheet feature, this is a part the documentation frameworks that are integrated in STACKn.

proposed by Sommerville (2016, p. 102). Sommerville describes these concepts according to the following: User requirements are a *"statement of what services the system is expected to provide to system users and the constraints under which it must operate"*, and System requirements are *"descriptions of the software system's functions, services, and operational constraints. The system requirements document [...] should define exactly what is to be implemented."* The requirements can be seen in Figure 6.

5.3 Architectural Design

Before implementing reproducibility features in STACKn, a necessary step was to define the design of the architecture as this is the first stage in the software design process according to Sommerville (2016, p. 168). Since our work for this book revolved around adding functionality to an already existing and fairly complex system, any architectural decisions had to be made with STACKn's existing structure and architecture in mind. Whether it was about adding functionality to the CLI, the user interface of STACKn Studio, or any back-end solutions, we examined how existing functionality of these parts of the system was already implemented.

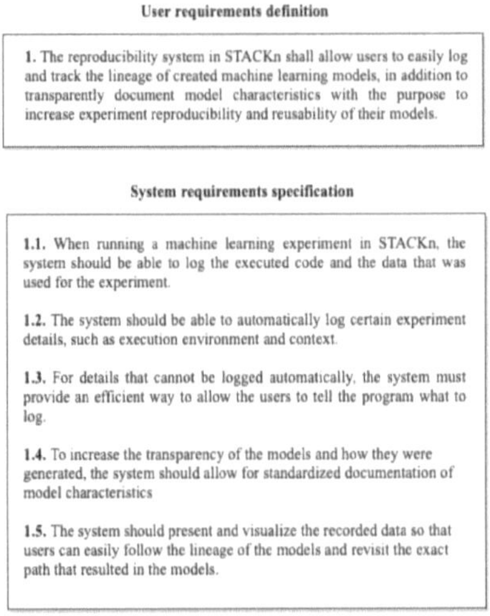

User requirements definition

1. The reproducibility system in STACKn shall allow users to easily log and track the lineage of created machine learning models, in addition to transparently document model characteristics with the purpose to increase experiment reproducibility and reusability of their models.

System requirements specification

1.1. When running a machine learning experiment in STACKn, the system should be able to log the executed code and the data that was used for the experiment.

1.2. The system should be able to automatically log certain experiment details, such as execution environment and context.

1.3. For details that cannot be logged automatically, the system must provide an efficient way to allow the users to tell the program what to log.

1.4. To increase the transparency of the models and how they were generated, the system should allow for standardized documentation of model characteristics

1.5. The system should present and visualize the recorded data so that users can easily follow the lineage of the models and revisit the exact path that resulted in the models.

Figure 6: *User requirements and System requirements identified for the reproducibility system in STACKn. User requirements are the services that the system should offer users, and System requirements are the functions, services and constraints that define exactly what should be implemented. Based on definitions provided by Sommerville (2016).*

Using an approach to architectural design as proposed by Sommerville (2016), the architecture type most in line with how we expected to build the reproducibility features was an *Application architecture*. More specifically, we designed the architecture based on a *Transaction processing application*, which are *"designed to process user requests for information from a database, or requests to update a database"* according to Sommerville (2016, p. 186). To ensure the provenance of models in STACKn and safe storage of all components that make it possible to the long-term reproducibility of experiments, information had to be stored in a database. The point is that this should make it easier for users to go back in time and see exactly how experiments have been conducted, as opposed to, for example, scenarios in which the user documents manually in a text document on the computer that easily gets after a certain amount of time. This is why we based our architecture on the Transaction processing application design; see Figure 7 for a proposed design of the architecture.

5.4   Design and Implementation

Here, we describe how we implemented our reproducibility features and how they work. The components and features described below are not presented in the order that they were developed during this project; rather, the presentation follows, in broad strokes, a chronological structure according to how the components are connected in the flow of the code and in line with how data is passed between components when running an experiment in STACKn.

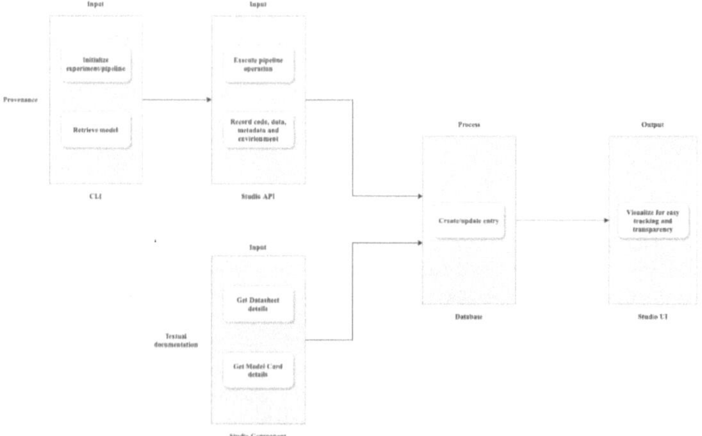

Figure 7: *Illustration of the proposed design of the architecture to the system that should increase the support for reproducibility in STACKn. Based on the architecture for Transaction processing applications as presented in Figure 6.16 in Sommerville (2016).*

5.4.1 STACKn CLI

Two new commands have been added to the CLI in STACKn – *stackn dvc* and *stackn run*, both with their own purposes. We explain the functions of these here. Everything we present here is implemented using the Python language.

Command 1: "stackn dvc"
Data Version Control, or DVC, is a management tool for large files such as ML experiments or datasets. Usually in software development, Git is used to manage different versions of the code itself. However, when it comes to really large dataset files, pushing them to a GitHub repository for version control is no longer an option. With DVC, the user can tag large files with a, only a few bytes big, .dvc-file, which contains a unique tag pointing to the location of the actual file, which in most cases would be a cloud storage. This small file can then, instead of the whole dataset or experiment file, be pushed to a Git repository.

In the STACKn CLI, we have implemented a new command; stackn dvc. This command sets up a DVC repository in the current directory with a remote store automatically set to the MinIO-bucket "datasets" of the current STACKn project. From that point on, the user can utilize the full array of DVC-commands available. See the DVC documentation[18] for more information.

Using these commands, a user that works on a particular STACKn project can generate a .dvc file for each dataset that are used to conduct an experiment, whether it be the dataset used for training, validation or testing of a machine learning model. When the user then makes a Git commit for the code that corresponds to the experiment, the .dvc files that represent the versions of the used datasets are captured in the Git commit. Therefore, the GitHub repository that is used for the current project can also be seen as a kind of data repository in which the researchers not only share their experiment code, but also the versions of the analysis data that they use for model experiments.

Command 2: "stackn run"

This command is connected to the Studio API – the component of STACKn that manages requests between the client and the database server – and is used to track and store information that is necessary for reproducibility in the database. Arguably one of the most important components in the context of our purpose, if not the most important component, this command was developed with the purpose to execute a sequence of machine learning operations (experiment) as determined by the user, and log and store details about each conducted experiment. When using this command, users must first determine which operations that should constitute the experiment; that is, which files of code that should be executed in sequence when running the command. For example, when users have written code for data pre-processing, model training, model testing and model predictions, they can tell the program to execute this simple pipeline by assigning this information in a configuration file. For this project, we made the configuration file as a simple JSON object, see example code below. Note that the entry on line 2 tells the program to install all pip dependencies that have been defined in the project's requirements.txt file. Lines 4-7 define the files of code that should be executed when calling stackn run.

```
1  {
2      "requirements": true,
3      "pipeline": {
4          "load_convert": "src/data/load_convert.py",
5          "reshape_normalize": "src/data/reshape_normalize.py",
6          "train": "src/models/train.py",
7          "predict": "src/models/predict.py"
8      }
9  }
```

When running this command, the user has the possibility pass a boolean variable -log-off to tell the program to not log any information about the executed experiment. This could, for example, be useful in minor experiments where logging would not be necessary. If data should be logged, the user must tell the program which STACKn model that should be the focus of the experiment. Note that in order for this to work, the user must have created the model using the API of STACKn and it needs to exist in the database. If the model does not exist in the Studio database, the program raises an error and will not continue.

Finally, the user can pass parameters telling the program which ID that should be assigned to the experiment; this parameter is not required for the user to define, however, and defaults to a random string of 32 characters, both digits and letters. The ID that is assigned to the experiment is crucial for the entire logging procedure, as it is the single unique identifier that ties the specific run to the tracked data.

To illustrate, suppose the user wants to experiment with a Convolutional Neural Network model for natural language processing tasks that has been created in STACKn under the generic name *NLP-model* (further specifics on the model is irrelevant here). Also assume this user has defined the configuration of the pipeline according to the JSON object seen in the example above. In this case, the CLI command to run the pipeline would simply look like this:

```
stackn run -m NLP-model
```

Running the above command, the user tells the program to pre-process the analysis data according to `load_convert.py` and `reshape_normalize.py`, followed by model training and model prediction according to `train.py` and `predict.py`, respectively. The logic of `stackn run` also captures relevant data for storage in Studio. If the user simply wants to train the model without logging any data, the user would instead run the following command:

```
stackn run --log-off
```

Simply running the CLI command `stackn run` will, if the logging is activated, automatically extract details about the context in which the user's machine learning model is being run. This includes the user's Python version, details about the local machine's system and CPU, and the source code. The source code is represented by the hash code of the latest Git commit in the Git repository where the user was operating when conducting the experiment. As such, the program raises a warning and pause the execution of the code if the user would start an experiment in a non-Git repository or (in cases where the user is in a Git repository) if there are uncommitted files in the user's working branch of the repository. If an experiment is initialized in a non-Git repository, the user is prompted to either 1) continue the experiment in this repository, or 2) exit the program and move to an active Git repository before running `stackn run` again. If an experiment is initialized in a Git repository in which there are uncommitted changes, the user is prompted to either 1) continue the experiment with uncommitted files, or 2) commit the file changes before running `stackn run` again.

In both scenarios described in the paragraph above, alternative 2) is clearly the recommended choice if users strive to increase the future reproducibility of their experiments; following recommendations by Stodden & Miguez (2014), code references through Git are needed for appropriate code versioning and reproducibility of experiments. As such, it is important that the program is clear on this note and tries to push the user in the right direction: For example, the program clearly states in the command line which option that is recommended for the purpose of high reproducibility.

5.4.2 Tracking Client

As described in 5.4.1, some information about the user's model experiments is automatically generated and logged when the user calls `stackn run`, such as code version (most recent Git commit) and hardware (such as CPU specifications). However, the automatic logging solution described in 5.4.1 cannot be used "as is" for other important metadata – such as hyperparameters and performance metrics – that are crucial to isolate and store to ensure reproducibility of model experiments. While the most ideal scenario would allow for automatic logging of such metadata, it would have been too complex to implement a fully automatic solution with the time resources available for this project. It is also the case that automatic logging would require some kind of manual input from the user in most cases (such as the users being required to name their variables in a specific way); as such, it could be argued that fully automatic logging is not even possible.

To enable the users of STACKn to log and track metadata from a model experiment, a tracking device called *Tracking Client* was created to facilitate semi-automatic logging of different model data. The Tracking Client is a Python class that contains four methods that are used for logging of data that is generated or defined during model experiments; the functions are described in the list below. Note that the user is not required to use these functions, but it is recommended in order to promote experiment reproducibility. Also, while the user must manually choose which data to store and explicitly tell the tracking client to store them, one could argue that it is a redundant feature. We argue that this is not the case as the purpose of the Tracking Client, albeit simple in nature, is to make the process of storing data more intuitive and organized for the user and facilitate central storage of important data. Making the Tracking Client simple in nature also has its benefits, as it makes it easier for users to understand. In short, the Tracking Client has the main responsibility of handling *metadata* and other key *artifacts* that are generated during model experiments.

- `log_params()`: Method used for storing parameters that are important for the model behavior, such that hyperparameters for controlling the model training.

- `log_metrics()`: Method used for storing metrics, performance measures or other quantifiable variables indicating how well the model performs.

- `log_model()`: Method used for storing information about the trained model. This includes the model type and the trained model object, typically a JSON object when using Keras to build models.

- `save_tracking()`: Method used for saving the logged data as a PKL file in the initialized project structure directory on the user's local machine.

The Tracking Client component was highly influenced by the solution seen in the MLflow platform, which also employs methods called `log_params` and `log_metrics`. The user can import the Tracking Client as a module in all files in the project structure that was initialized with the command `stackn init`. They do so by instantiating a class of the client to create an object that works as a kind of data processor that manages the data so that the Studio API in STACKn is able to later recognize which data to retrieve from the user's local machine for storage in the central database. After a Tracking Client object has been created, the user can take advantage of the tracking methods described above as they see fit. For example, the user can store training or validation accuracy for a model with the `log_metric()` method. When using the Tracking Client, it is important that the user labels the tracked metadata to make it easy to identify from which pipeline operation that generated the data. For example, if the user logs validation loss when validating a model, it should be labelled something like "Validation loss". Otherwise, it might be confused with the model loss metric value that was calculated when training the model.

5.4.3 Studio API

As the Studio API (Studio Client) was an existing part of STACKn, this was not created from the ground up as part of this project. Rather, modifications were made to the Studio API so that the data generated and logged during model training could be sent to the database on the server side of STACKn. When using the newly implemented CLI commands that we described in Section 5.4.1, it is the Studio API that performs the logic behind the commands. This includes executing all the scripts that are defined in the experiment configuration file `pipeline.json` and sending the metadata that is logged during model experiments to the database. Note that, similar to the two sections above, everything we implemented in Studio API was written in Python.

5.4.4 Database

Ideally, the data that is generated during experiments need to be stored in a resilient database on a central server instead of in a local storage on the user's machine; this would ensure that the data would be stored safely and that it would be out of harm's way in case of unexpected errors, such as data corruption, on the user's machine. At the same time, there was no need to incorporate a complex solution that would demand too much time of the development process, as this would be out of scope for this book paper.

First, the idea was to set up a MongoDB database by adding a helm chart that would bootstrap a MongoDB deployment on top of the existing Kubernetes cluster. As such, the database server would be created automatically when installing the STACKn deployment, and available for use inside the Kubernetes cluster in which STACKn was deployed. While this solution would have been optimal, there was a substantial risk and fear that efforts to make it work would have taken up too much of the time resources available for this project. The helm chart for MongoDB was incorporated into STACKn successfully, but the art of actually making it communicate with the Kubernetes cluster was not as trivial. The purpose and direction of this book paper is not focused on database design, but rather machine learning reproducibility; therefore, spending too much time on the complexities of designing a database for metadata storage would have been too much of a sidetrack to be justified.

Hence, in order to stay true to the purpose and research questions in this book paper, a different direction was chosen since optimizing the application's database management was not urgent for the goal of this project. As the API of the STACKn Studio component is built on the Django REST framework, it was possible to reuse components in the API and extend it for the purpose of metadata storage. When using Django to build web APIs, it is possible to define and use so called Django *Models*[19] for mapping data to a database storage. Models define a data object and what kind of information to store and the eventual relationship between database entries. For our purposes, we created one model for experiment logs with general details about an experiment and one model containing information about the metadata that is logged with the Tracking Client, and also one model each for the Datasheet and Model Card features that are explained in Sections 5.4.7 and 5.4.6, respectively.

5.4.5 STACKn Studio

As previously mentioned, we have implemented features that tracks and stores information about model experiments in STACKn's database. However, this information needed to be shown to the user in some way to make them valuable. This is where the STACKn Studio comes in. As with the Studio API, the UI of STACKn Studio was already in place prior to our work, and any changes we made here demanded careful consideration as not to disrupt the existing design and workflows. Mainly two parts of STACKn Studio have been altered with the introduction of the new reproducibility functionalities; The "Model Details" and "Dataset" views inside a STACKn Project. The implementation of these new views is done primarily using HTML, with logic provided through Django *Views*[20] that takes web requests and returns web responses. The Django integration comes into use when any information stored in the Django database is to be rendered, such as lists of models or datasets, any logged artifacts about a model, as well as Model Cards and Datasheets.

Each model in STACKn has a view with "Model Details", where details about the model is displayed. Several different categories of model information is available through the use of tabs at the top part of the view. This tab bar came across as the natural place to include information that is generated when calling `stackn run`, such as logged model parameters, performance metrics, as well as any user generated Model Card. In total, three new tabs have been added: "Logs", "Performance" and "Model Card". Under "Logs", a list with information of all the conducted experiments for the current model is shown, with the option to show additional information such as a complete list of logged metadata and artifacts for a specific experiment. The "Performance" tab, which was mainly created using JavaScript, displays a graph that plots how performance metrics have changed over time; the performance metric that the user is interested in is chosen in a drop-down menu next to the graph. The metric is shown on the Y-axis, while the X-axis represents time measured in days. While a graph showing the evolution of model performance might not be directly related to experiment reproducibility, this was a common feature in many existing ML platforms. In both the "Logs" and "Performance" tab, we added an "introduction box" at the top of the view where we included a short description of what each view presented and what the users could do in each view.

The tab labeled "Model Card" is where the user can view or fill out a Model Card for the currently selected model; the contents of this tab is described in the upcoming section.

5.4.6 STACKn Model Cards

Staying on the "Model Details" view, the third tab that we implemented was the one for creating a new instance or inspecting an existing Model Card. Similar to the other Studio features, the Model Card was created using HTML and Django Views logic, with the difference that we here also employed the concept of Django *Forms*[21] that can be used to accept input from the users. The idea was first to integrate Model Cards through the Model Card Toolkit mentioned in Section 4.2, but we chose to implement it ourselves. We based our solution on the existing toolkit and adopted the concepts suggested by Mitchell et al. (2019), namely to have a number of sections presenting details about model characteristics and performance.

Considering that many users might not know what the concept of Model Cards is about, we included an "introduction box" here as well, with explanations to what a Model Card is and with a reference to the paper by Mitchell et al. (2019). We included a clear button which the user can click to be redirected to a view where the Forms are shown. The Forms were divided into sections and subsections, according to what is presented by Mitchell et al. (2019, pp. 3–6), and we included explanations as to what information each Form should contain. When the user is happy with the information in the Model Card, the user can create the card and is redirected back to the "Model Details" view. When the Model Card is submitted, the user-input in the Forms is used to populate a Django Model, appropriately named "ModelCard", and saved in the database. When the user navigates back to the "Model Card" tab, the created Model Card is rendered and the information is presented in the correct sections.

Due to the time constraints of this book project, all sections as suggested by Mitchell et al. (2019) are not included in the Model Card. All sections, except the ones for "Evaluation Data", "Training Data" and "Quantitative Analyses", are included. However, the sections related to data are not considered necessary as these are covered in the feature that is explained next, namely Datasheets.

5.4.7 STACKn Datasheets

When the user navigates to the "Datasets" view in Studio, the user can upload files to MinIO from this view, specifically to a MinIO bucket called "datasets". Subsequently, the uploaded data file is shown as a dataset in STACKn Studio under the "Datasets" view, including a table of all objects currently uploaded in the MinIO-bucket "datasets". In the table, we created a new column labelled "Datasheet" containing a button for each dataset in the list. When clicking on the Datasheet button for a dataset, the user is redirected to a view where they can, depending on whether the button says "Create" or "Show", populate or investigate a Datasheet for the selected dataset. We based this feature on the same idea as the Model Cards, namely Django Forms. This means that when users have populated a Datasheet using the predetermined Forms, a Django Model called "Dataset" is initiated with a field "Datasheet" which is populated with the provided answers from the Form. As such, we implementation of Model Cards and Datasheets was basically the same with regards to technicality.

Note that a fully fledged d atasheet a s s uggested b y G ebru e t a l. (2018) h as a lmost sixty different questions that is to be answered by the creator of the dataset; while a datasheet with all of them answered would be optimal in terms of transparency and reproducibility, it is unlikely that STACKn users would go through all the hassle to fill out a complete d atasheet. In light of this, we hand-picked nine questions from the original 59 that we believe are the most important. In addition to these, one question was formulated by us, namely "For what purpose, and by whom is the dataset used?". The set of questions is easily expanded or reduced, since the Django view is rendered dynamically using a list of questions available in a Python file. If questions are to be added or removed, said list is the only thing that need to be edited in the code. The nine questions presently selected from the original datasheet questions are as follows:

- Who created the dataset (e.g., which team, research group) and on behalf of which entity (e.g., company, institution, organization)?

- What do the instances that comprise the dataset represent (e.g., documents, photos, people, countries)? Are there multiple types of instances (e.g., movies, users, and ratings; people and interactions between them; nodes and edges)? Please provide a description.

- What data does each instance consist of? "Raw" data (e.g., unprocessed text or images) or features? In either case, please provide a description.

- Are there any errors, sources of noise, or redundancies in the dataset? If so, please provide a description.

- Does the dataset contain data that might be considered confidential (e.g., data that is protected by legal privilege or by doctor-patient confidentiality, data that includes the content of individuals' non-public communications)? If so, please provide a description.

- What mechanisms or procedures were used to collect the data (e.g., hardware apparatus or sensor, manual human curation, soft-ware program, software API)? How were these mechanisms or procedures validated?

- Were any ethical review processes conducted (e.g., by an institutional review board)? If so, please provide a description of these review processes, including the outcomes, as well as a link or other access point to any supporting documentation.

- How can the owner/curator/manager of the dataset be contacted (e.g., email address)?

- Any other comments?

5.5 Software Testing

As part of the software engineering process, we conducted testing of the created software continuously throughout the entire project. As proposed by Sommerville (2016, p. 231), the software testing stage consists of three steps: 1) *Development testing*, 2) *Release testing* and 3) *User testing*. Sommerville states that a commercial software system usually has to go through all three stages before releasing it, but due to the time constraints imposed on us during this project, the only testing stage that we managed to do was Development testing. This testing stage is conducted in parallel to the development phases to discover bugs and defects. This was done regularly by us so to keep the system updated constantly and to make sure that it would never fail at later stages of the development.

The Development testing consisted of three phases, as suggested by Sommerville (2016, p. 232): 1) *Unit testing, Component testing*, and 3) *System testing*. All of these parts were carried out by us continuously in the following manner: Whenever a new class or method was implemented – for example the Tracking Client as described in Section 5.4.2 – this single piece of code was tested through Unit testing; in the case of the Tracking Client, this was done by checking if its tracking methods worked as intended by creating shorter scripts of testing code and isolate it from the other parts of STACKn. When this was completed, this unit was integrated with other components in the system and tested through Component testing; in the case of the Tracking Client, this consisted of integrating it in the STACKn project structure and test if it was able to communicate with the Studio API. And finally, we conducted System testing to see if this unit was able to be integrated and work with the entire STACKn structure; in the case of the Tracking Client, this consisted of testing whether the logged data could be stored and sent all the way to the database, and subsequently presented and visualized as intended in the UI of STACKn Studio.

6 Evaluation

This section of the paper is devoted to a thorough evaluation of the implemented solution, with the purpose to make an objective measurement of the degree of reproducibility that is supported after integrating our reproducibility features. This section is organized according to the following structure: First, the newly implemented features is demonstrated to give an overview of how they work and why they contribute to the reproducibility and transparency of machine learning experiments, which is important to understand when going through the evaluation. For clarity, *Provenance features* are here on out used to refer to every newly implemented feature related to the CLI commands `stackn dvc`, `stackn run` and the Tracking Client; *Documentation frameworks* is here on out used to referr to the Datasheet and Model Card features.

Following the demonstration, we present the evaluation method we use to objectively quantify the degree of experimental reproducibility that STACKn supports; for this purpose, we employ a method for calculating the the degree of reproducibility in ML platforms as presented by Isdahl & Gundersen (2019). After this, we present an analysis model that is used to analyze the degree of reproducibility in STACKn from the perspective of the spectrum of reproducibility explained in Section 1.1.1, mainly with the purpose to answer the final research question stated earlier. Finally, we present the evaluation and the results thereof.

6.1 Reproducibility Features in STACKn

Here, we demonstrate the newly implemented reproducibility features in STACKn. This is mainly done with the purpose to showcase how the different features work in practice and to provide explanations as to how the features contribute to the reproducibility of machine learning experiments that are conducted in STACKn. We demonstrate each feature type separately, starting with the provenance features integrated through `stackn run`, `stackn dvc` and the Tracking Client, followed by a demonstration of the documentation frameworks that consist of the Datasheet and Model Card features. For a summary of the "reproducibility workflow" that we claim that these features support and that we attempt to showcase in this section, see Figure 20 in Section 6.5.

Note that we have not chosen to use this experimental approach to evaluate the degree of reproducibility that is supported in STACKn. Sandve et al. (2013) note that a minimal requirement for reproducibility is that researchers should be able to at least reproduce their own results. Although they note that this requirement might be quite weak, it still demands a substantial level of care to details in order to be met, at least in computational science related experiments in which there is a vast number of variables, components and steps that must be considered when documenting the research process. This is basically what we do in this section to demonstrate the reproducibility, so it is not irrelevant to make our own experiments to test the reproducibility in STACKn. However, it is worth noting that Gundersen & Kjensmo (2018) define this process as repeatability, rather than reproducibility: They argue that the experiments must be conducted and reproduced by an independent person or group in order to label an experiment as reproducible.

As we conduct simple experiments ourselves to showcase the implemented features, what is presented in this section is primarily be a demonstration of the aforementioned features and not an objective evaluation of how the experimental reproducibility is affected in STACKn by these features. If we would use this method to evaluate the degree of reproducibility of experiments in STACKn, one could argue that we would only be testing for repeatability as opposed to reproducibility according to the definitions by Gundersen & Kjensmo (2018), which would weaken the validity of our conclusions. Clearly, the optimal scenario would be to let an independent person or group test our solutions if we had chosen to take an experimental approach to measure the degree of reproducibility that STACKn supports. That is why we have used a different method to achieve a more objective evaluation of how our solutions affect the degree of experimental reproducibility in STACKn, see Sections 6.2 and 6.4. In short, this section does not present the official evaluation of the impact of our implemented features, but rather a simple demonstration of how they work and how they can be used for conducting reproducible experiments in STACKn. Also, a clear picture of how the implemented features work and how they contribute to reproducibility is necessary to have in order to understand the subsequent evaluation presented in Section 6.4.

6.1.1 Experimental Setup

The demonstration of our implemented reproducibility features in STACKn is carried out by conducting short and simple experiments with a convolutional neural network built with the Keras API. The model is tasked with correctly classifying hand written digits from the MNIST dataset[22]. Using this simple CNN as a starting point, we showcase how the provenance features in STACKn can be used to track changes that are made to the model between experiments. Changes are then made to model characteristics such as net structure, optimizer algorithm and activation functions, or training parameters such as batch size and number of epochs, after which the model is retrained. Furthermore, we showcase how the integrated documentation frameworks can be used to provide transparency to the model.

The code for building the initial model can be found in the STACKn tutorial documentation[23]. For reference, the sequential CNN model is initially trained with the Adadelta optimizer algorithm, a batch size of 32 and one epoch. Furthermore, the initial sequence of model layers is as follows, with eight layers in total:

- 2D convolutional layer with input tensor of shape [28,28,1], kernel size [3,3], activation function ReLu

- 2D convolutional layer with kernel size [3,3], activation function ReLu

- Max-Pooling layer with pool size [2,2]

- Dropout layer with dropout rate 0.25

- Flatten layer for converting the input to a 1-dimensional array

- Dense layer with dimensionality of 128 units, activation function ReLu

- Dropout layer with dropout rate 0.5

- Dense layer with dimensionality of 10 units, activation function Softmax

The exact changes to the model are Ad hoc and might be of a highly experimental nature and may not be grounded in theoretical explanations, the important thing is that clear changes in the results can be seen between experiments. As we primarily conduct this experiment to demonstrate our solutions, all of the exact changes is not motivated in detail. As such, some decisions we make with regards to experiment parameters and setup might seem random or arbitrary in true "trial and error" fashion.

In contrast to the models that many of the STACKn users will most likely be working with, the model used for this experiment is very simple and easy to work with, and might not make sense to deploy in a production context. As we want to demonstrate how our newly implemented features can be used to track the lineage of models and transparent documentation of the model characteristics, we did not find it necessary to spend time on building a complex and optimized model from scratch. Simply put, it does not make sense in the context of this paper to spend time to tune and optimize hyperparameters to try to create a state-of-the-art model; this would not have been relevant for our demonstration purposes. The important concept to showcase is simply how model changes can be traced and then revisited with the purpose to reproduce or reuse in the exact same manner as in the original experiment. We argue that the principles and basic ideas of our solutions are the same for all models in STACKn, regardless of complexity, which is why we felt confident in our decision to demonstrate with a simple model that we did not build from the ground up.

Before starting the demonstration it is worth noting that we created a new Git repository called MNIST where the experiment code is recorded. It is important that the user is working in a Git project since the reproducibility of each conducted experiment in STACKn depends a lot on whether the user works in a Git project or locally. Furthermore, an empty project structure was initialized using `stackn init` and a STACKn Project called MNIST was created in the Studio UI for this demonstration which was set to active using the CLI command `stackn set project -p MNIST`. An initial model called "mnist" was created with the CLI command `stackn create model -n mnist -r minor`; this generated a model of version v0.1.0 in the Studio UI and new Kubernetes resources were initialized for the model.

6.1.2 Provenance Features

Here, we present the provenance features that we implemented in STACKn and explain how they contribute to reproducibility through lineage tracking and reusability of machine learning models. To do this, we first give a general overview of how the features are used, followed by a presentation of two examples showcasing scenarios where these features can be used for reproducibility and reusability.

The first step to do when conducting a machine learning experiment in STACKn with our newly implemented provenance features is to setup a DVC configuration file with the correct AWS credentials for the currently active project, making it possible to store experiment data to the remote storage MinIO that is unique for the active project. By running the newly implemented CLI command `stackn dvc`, a .dvc directory is created in the initialized project structure with correct configuration for the remote cloud storage where we store copies of the data that we are using.

After this, the next step is to retrieve the analysis data, both training and test data. The raw data is then processed, first by converting the data to numpy arrays to create the interim data, then reshaping and normalizing the data to its final form. To track the data and any eventual changes to it, the DVC command `dvc add` is used to start tracking of the raw, interim, and processed data. Information about each dataset is stored in separate .dvc files that are versioned using Git, and we do not have to push the real datasets to the Git repository; instead, if the real data is stored in MinIO, we only need to store the .dvc files in the Git repository in order to retrieve the data that was used for the experiment at later stages when the experiment should be reproduced. As such, we can add the data to a .gitignore file, assuming we make sure that we commit the .dvc files every time changes are made to the datasets.

We then add the code for training the model with the processed data. Using the Tracking Client and calling the `stackn run` command to execute the experiment, general information about the experiment, such as code version and hardware specifications, is logged in the Studio database and shown in the UI, in addition to interesting metadata and artifacts that is generated and logged by the user. Consequently, we make some changes to the experiment setup to modify the model performance, once again using the Tracking Client to log selected metadata and artifacts. After running four experiments with different setup of hyperparameters, we first navigate to the newly implemented view that shows a list of experiment logs that are generated automatically when calling `stackn run`, seen in Figure 8. This view presents the user with a list containing details for all experiments that have been conducted with `stackn run` for the model "mnist" version v0.1.0. As seen in this list, each experiment entry is labelled with a unique ID (truncated format), date and time for when the experiment was conducted, the STACKn user who conducted the experiment, and two clickable links that activate pop-up windows for both general information about the experiment and metadata and artifacts that have been logged with the Tracking Client. The most recent experiment is shown at the top of the list, as can be seen in the column for data and time.

Now, we use the initial experiment with ID starting with "16a2e30e565811..." as seen in the log list to showcase the experiment details that are recorded when calling `stackn run`. If we click on "Show info" under the column "General info" for this experiment, a pop-up window is activated that shows general details that are automatically generated when conducting the experiment. These general details include the exact ID for the experiment in its original format, date and time for execution for the experiment, the Git repository where the user operated when running the experiment, the Git commit hash for the executed code version, and system information and hardware specifications for the machine that was used to conduct the experiment. General details for the experiment used as an example here can be seen in Figure 9.

If we close this pop-up window and click on "Show metadata" under the column "Metadata" for this experiment, another pop-up window is activated; this modal contains any eventual metadata and artifacts that were logged during the experiment with the Tracking Client. In this case, we have logged the batch size and number of epochs used for model training, the resulting training accuracy, categorical crossentropy loss and mean-squared error as model performance indicators, and model details. The model details include model type and model structure, and this is shown as a collapsible HTML element due to the quantity of information in the model JSON. Metadata and artifacts for the example experiment can be seen in Figure 10. Note that Figure 10 does not show the logged model details as this information is hidden through the collapsible web element; instead, see Figure 11 where the button "Model details" have been pressed and the logged model details are shown.

If we then navigate to the newly implemented view under the tab labelled "Performance", we can plot the performance metrics that we have logged during our model experiments. In the dropdown list seen above the plot area, we can select training accuracy to see how this metric varies over time. An example of such a plot can be seen in Figure 12; in this case, training accuracy is shown on the Y-axis, while the X-axis shows time measured in days. If we hover over the data points in the plot, a box is shown with information about the specific experiment; this includes

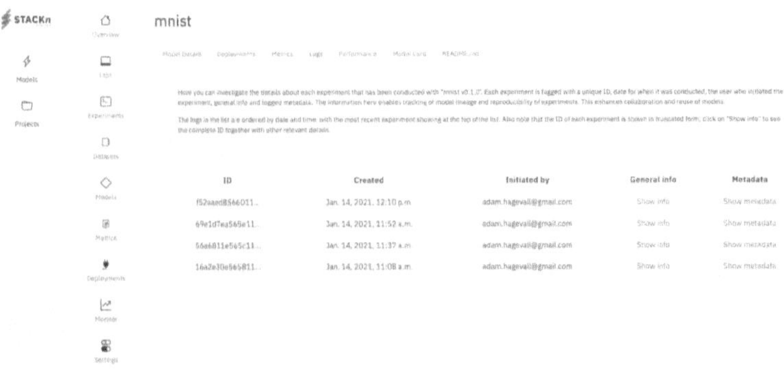

Figure 8: *A newly implemented view in the UI for STACKn Studio, seen in the details page for the model "mnist". This view shows a list of logs containing information about conducted experiments. This view shows the unique ID for the conducted experiment, date and time for when the experiment run was initiated, which user that initiated the experiment, and two clickable links that activate pop-up windows for both general information about the experiment and tracked metadata and artifacts.*

the experiment's ID, date and time for the experiment, and the parameters that were logged for this experiment. This kind of plot provides a good overview of how the model performance has changed over time, and can be used to quickly determine what experiments that generated good models. It can also provide a quick overview of what learning parameters that gave rise to the exact models. Note that the metrics that can be selected for this kind of visual presentation are only those that are measured as single-value integers.

What follows are two examples of scenarios where these provenance features can be used for reproducibility purposes. The first example simply showcases how these features can be used to reproduce the results exactly as they are in the original experiments, in this case with the purpose to validate the results. The second example showcases how a model can be reused in the future. The examples are not entirely different from each other, but still highlights tow different use cases where the provenance features can be used.

Example 1: Reproduce results exactly for validity
Here, assume an independent researcher investigates the validity of the experiment results shown in Figure 12. For this purpose, the user can first go back to the list of logs in STACKn and check out the exact code for each experiment and use DVC to get the data that was used for the

43

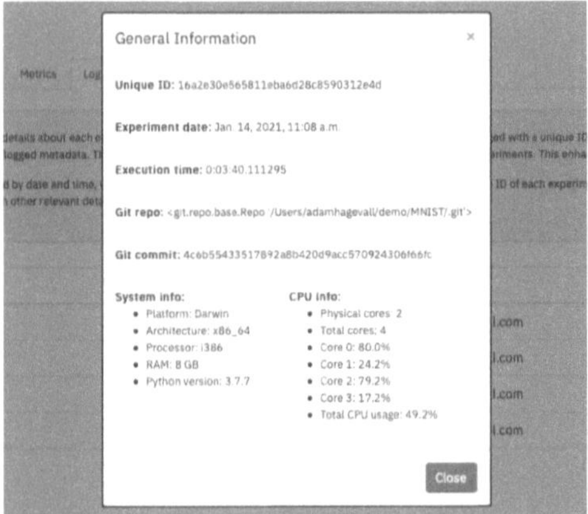

Figure 9: *General details about the conducted experiment. This modal window shows complete experiment ID, date and time, experiment execution time, Git details, system info and hardware specifications. Note that "Git repo" is presented as the full path on the local system; since the .git folder lies in MNIST, this is the name of the actual Git repository on GitHub.*

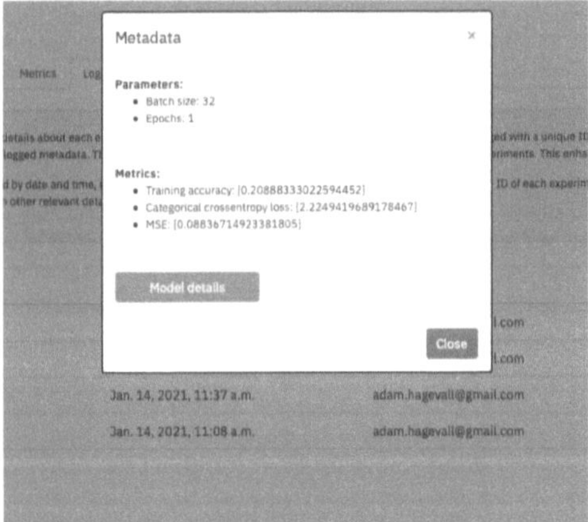

Figure 10: *Experiment metadata and artifacts logged with the Tracking Client when running the CLI command stackn run. Here, we have logged batch size and number of epochs used for training, and the resulting training accuracy, categorical crossentropy loss and mean-squared error. Model details have also been logged, but they are hidden through a collapsible HTML element.*

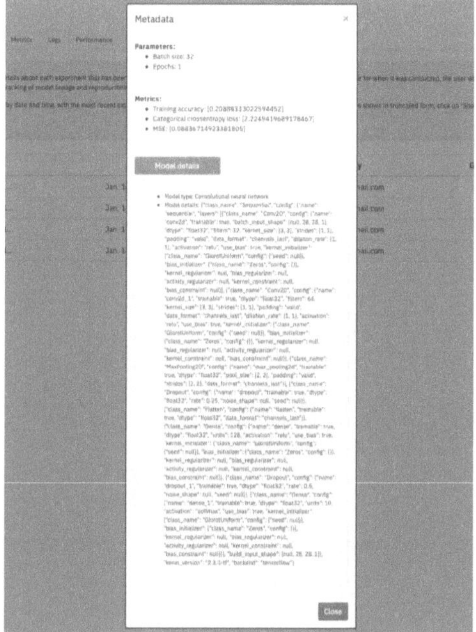

Figure 11: *After clicking on "Show details" in the metadata window, the logged model details presents itself. This shows the model type and the model object presented as the JSON that was created after training the model with Keras. The model object shows the exact model layers and layer configurations (activation functions and layer type, and so on); while it might be difficult to see the exact specification of the model object due to the deeply embedded JSON structure of the model object, it demonstrates how the Tracking Client can be used to record models when conducting experiments in STACKn.*

experiment, and then simply call `stackn run -m mnist`. After doing this, the plot is updated with new experiment runs and can be seen in Figure 13. As can be seen, the new experiments have generated the exact same results as in the original experiments; while small deviations can be seen in the results, this is in accordance with statements by Gundersen & Kjensmo (2018) who say that any differences in the achieved results must be due to differences in hardware.

A more realistic scenario of how this can be used to validate results through reproduction could be as follows. Assume the performance plot indicates a model with a near perfect training accuracy. Suppose an independent user is skeptical of these results and wonders whether the model has been evaluated on a validation dataset. This user can then check out the exact code that produced the model (just like above, by investigating the log that corresponds to the experiment) and notices that the model in fact has not been tested. Using the same code and data as in the original experiment – enabled by our reproducibility features – this user can test the same

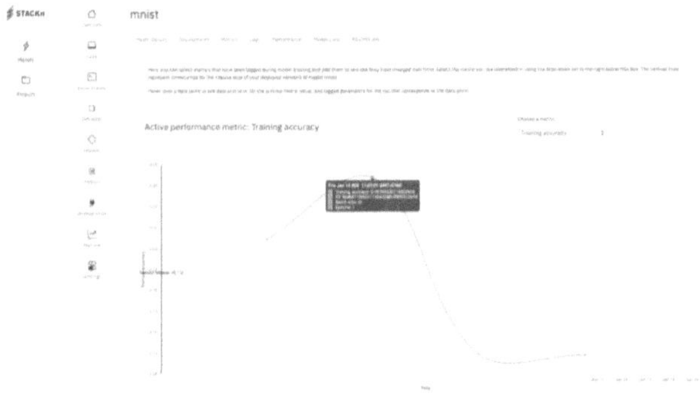

Figure 12: *Plot showing how a performance metric (in this case training accuracy) for the model has changed over time. Used for easy tracking of the model lineage in terms of how the performance evolves. All experiments were conducted on January 14, 2021.*

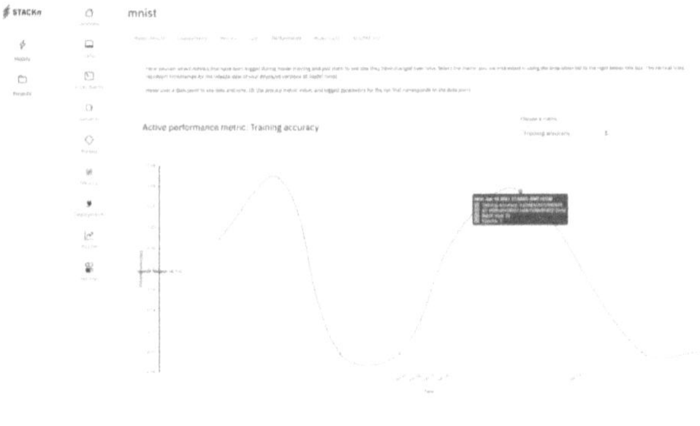

Figure 13: *Updated plot showing how the training accuracy for the first four experiments have been reproduced with our provenance features. The experiments were reproduced on January 18, 2021 and January 19, 2021, four and five days, respectively, after the original experiments were conducted.*

model on unseen data, and the results might show a test accuracy that is significantly lower than the training accuracy. This indicates that the model could have overfitted during training, and the user can warn of deploying this version of the model in a production setting as the model would probably be inappropriate to use. Of course, it does not need to be an independent user who conducts this kind of reproduction process, it could just as well be the original researcher if he/she realizes that he/she forgot to test the model when conducting the original experiment. This highlights the importance of actually being able to go back to previously conducted experiments and reproduce and validate results, a process our provenance features is able to aid in.

Example 2: Reuse model

Assume we feel satisfied with the model structure that was used for the first experiment in Figure 12 and we want to use this exact model structure when releasing version v0.2.0 of the model "mnist". For some reason, however, the training accuracy is quite low: Approximately 0.3. If we hover over the data point for this experiment, we quickly notice that only one epoch was used for model training; without our provenance features, this realization would probably have been much more difficult to reach, especially if this experiment had been conducted by another user.

Assume that we want to use a larger number of epochs (everything else being the same) to hopefully achieve a higher training and validation accuracy and a lower training and validation loss. We can then take the ID of this experiment and extract which code version that was executed for this experiment; once again, we do this by checking out the Git commit that was recorded in the correct experiment log. Now, we can experiment with the number of epochs to achieve the model results that satisfies our purposes. We see an increase in the performance of our model and create version v0.2.0 of the model "mnist", see Figure 14. Note that the same basic idea can used for other purposes as well; for example, we might want to use the same basic CNN structure but train it on handwritten letters instead of handwritten digits to create an alphabet recognition model.

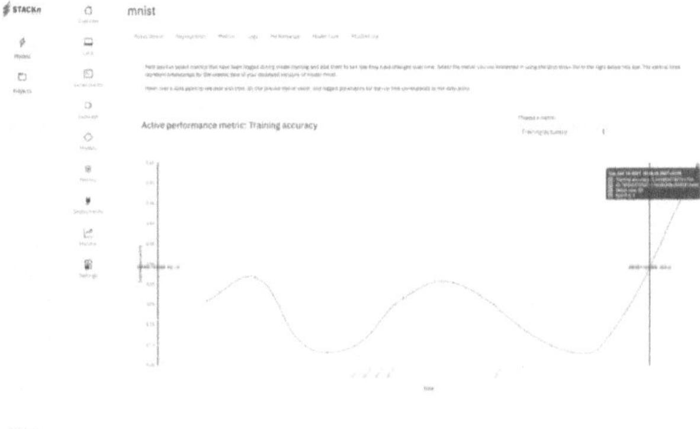

Figure 14: *Updated plot showing how version version v0.2.0 of the model "mnist" has been created after which the initial model structure was reused and retrained with a different number of epochs. The vertical line in the plot indicates when the new version of the model was created in STACKn.*

6.1.3   Documentation Frameworks

In the above examples, both the deployed model and the dataset used for training it was given supporting documentation using the Model Card and Datasheet, respectively. To illustrate how these supporting frameworks have been implemented, we illustrate how the GUI of STACKn has been altered through a series of images.

Before any model can be trained, access to training data is a necessity. In a STACKn project, each file uploaded to its corresponding MinIO-bucket "datasets", is be represented through a list in the STACKn GUI under the category *Datasets*. In Figure 15 we can see the datasets used for training and testing the MNIST model, as well as their raw and interim counterparts. One dataset, the unprocessed "train-images-idx3-ubyte", has been given a Datasheet; as indicated by the button for Datasheet for this particular dataset displaying "Show" instead of "Create".

Figure 15: *List of added datasets in the STACKn project MNIST, here seen with the option to create a Datasheet for the datasets ("Create") or investigate an existing Datasheet ("Show").*

If we then click on "Show", we are redirected to the actual Datasheet as filled out by the user, see Figure 16. In this view, the user can read information about the dataset, for instance what model that was trained with this dataset, or update the Datasheet if there are any details in the original Datasheet in need of modification. Furthermore, the user can also download the Datasheet as a PDF document in case the Datasheets needs to be distributed to individuals who does not use STACKn; for instance, this could include customers (assuming the creators of the particular model are business focused STACKn users) who uses models that are related to the particular data in a production setting.

If we then navigate to the detail page for the model "mnist" version v0.2.0 and move to the "Model Card" tab, we can create a Model Card for the model. An example Model Card can be seen in Figure 17, in which information about model characteristics and performance is shown. As can be seen, the Model Card is divided into six sections containing different kinds of information. Also shown is information about when the Model Card was created, which version of the Model Card that is currently showing, and when the Model Card was last updated. If a project member has not been initiated with the concept of Model Cards, there is an option to click on "What is a model card?" to show a description. Also, the user can update the Model Card by clicking on "Update model card" which is useful to do after conducting experiments with the model that might

Datasheet for train-images-idx3-ubyte

A datasheet is a standardized document used in order to provide information about the provenance, creation, and usage of a particular dataset. They serve to increase transparency and to make it easier for new users of the dataset to avoid unwanted biases. Below, a simplified version of a datasheet can be filled in. You can read more about Datasheets for Datasets at: https://arxiv.org/abs/1803.09010

Below you can see the answers provided on the datasheet for dataset train-images-idx3-ubyte. Press the button on the bottom of the page to update the datasheet.

For what purpose, and by whom is the dataset used?
Used for training of model MNIST v-0.1.0 by Adam Hagmark and Carl Wikström.

Who created the dataset (e.g., which team, research group) and on behalf of which entity (e.g., company, institution, organization)?
Yann LeCun, Courant Institute, NYU Corinna Cortes, Google Labs, New York Christopher J.C. Burges, Microsoft Research, Redmond

What do the instances that comprise the dataset represent (e.g., documents, photos, people, countries)? Are there multiple types of instances (e.g., movies, users, and ratings; people and interactions between them; nodes and edges)? Please provide a description.
Size-normalized and centered images of hand written digits.

What data does each instance consist of? "Raw" data (e.g., unprocessed text or images) or features, or has any preprocessing been done? In either case, please provide a description.
Size-normalization and centering in fixed-dimension images. No other preprocessing has been done on this particular dataset on our part. For the final dataset used to train the MNIST model, however: The dataset is first converted into numpy-arrays, which are reshaped to tensors of shape (28, 28, 1), and then converted to floats and it's RGB value is normalized to floats between 0 and 1.

Are there any errors, sources of noise, or redundancies in the dataset? If so, please provide a description.
Not that we know of

Figure 16: *An example of a Datasheet for the dataset "train-images-idx3-ubyte" which has been populated by the user. Here, the Datasheet is not shown in its entirety.*

inherently change the behavior of it, for example if the model has been trained with a different dataset. In this case, the user is redirected to a different view to update the Model Card, see Figure 18. This view is the same as the one users meet when creating the Model Card the first time, which contains guidelines to what kind of information that should be detailed in the card. Similar to Datasheets, users can also download Model Cards as a PDF document in case the Model Card needs to be distributed to individuals who does not use STACKn.

We argue that the addition of Datasheet and Model Card generation contributes greatly to the transparency of the models that are created in STACKn. For instance, users can take advantage of both of these features in scenarios similar to the one referred to as *Example 2* previously explained in Section 6.1.2. These documentation frameworks can be seen as improving the explainability of models as they provide details about their characteristics and how they are generated, providing sufficient amount of information for other users to determine if the models can be reused for their purposes. As such, they should provide sufficient information so that the models are not used in settings where they will create issues or for tasks that they are not trained for.

6.2 Quantifying Reproducibility in ML Platforms

In order to make an objective evaluation of the degree of reproducibility that STACKn supports, a structured method was needed to generate clear and quantifiable results. To quantify the degree of reproducibility in an objective way, we have taken the liberty of using the same evaluation method as presented by Isdahl & Gundersen (2019), which is an extension of the method proposed by Gundersen & Kjensmo (2018). The main difference between Isdahl & Kjensmo (2019) and Gundersen & Kjensmo (2018) is that the former conducted a survey of the degree of reproducibility found in 13 existing state-of-the-art machine learning platforms (some of which are presented in Sec-

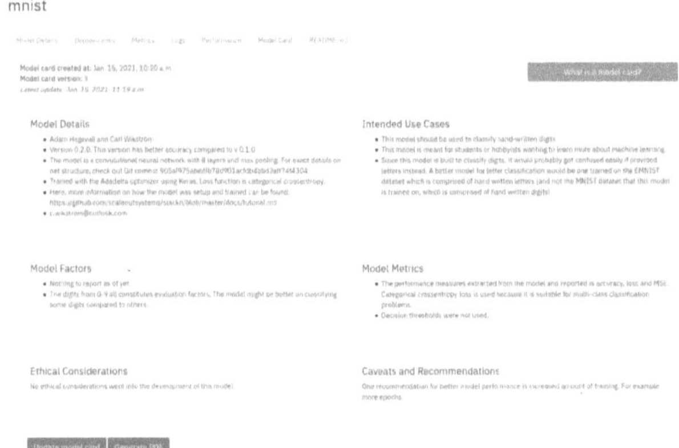

Figure 17: *An example of a Model Card for the model "mnist" version v0.2.0 which has been filled out by the user.*

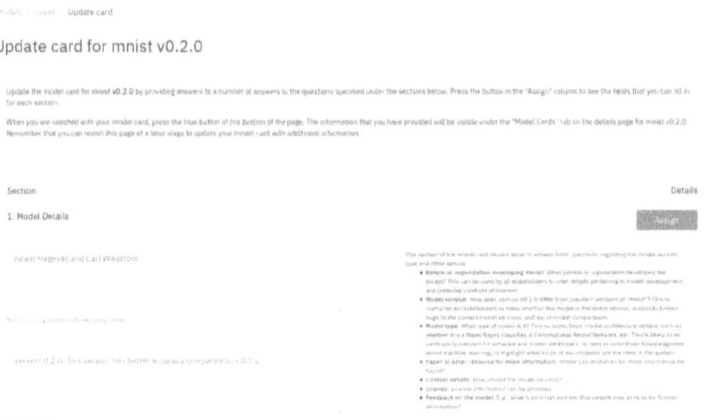

Figure 18: *The view in which the user can update the Model Card to reflect changes in the corresponding machine learning model. This view is the same as the one users meet when creating the Model Card the first time, which contains guidelines to what kind of information that should be detailed in the card. The blue box at the bottom right corner contains guidelines for how to provide information in the Model Card section called "Model Details".*

tion 4 of this paper), while the latter investigated the degree of reproducibility of results that had been presented in research papers at top AI conferences. Considering that STACKn is a machine learning platform similar to those evaluated by Isdahl & Gundersen (2019), we considered their method as an appropriate choice to use to determine how well STACKn supports reproducibility.

Furthermore, the decision to use a measurement method as created and presented by an independent party was taken with the purpose to reduce the bias towards our own implementation in STACKn: If we had used a method as defined by ourselves, the evaluation would likely risk leaning towards subjective assessment rather than objective, which would have opened up significantly for the conclusions of our evaluation to be disputed. To avoid bias from our side to the highest extent possible, we follow the methodology presented by Isdahl & Gundersen (2019) very closely and motivate our evaluation based on their reasoning when possible. Unfortunately, we believe that the evaluation is subject to some degree of bias despite of this, and we argue that it is unavoidable in this case because of the way that the evaluation method is structured (it is clear that some parts of the method demand some kind of subjective input). This becomes clear in later parts of the paper where it is discussed in more detail, but remember that the use of a third-party evaluation method was not meant to remove bias completely, but rather reduce it to a reasonable level where it cannot corrupt the results entirely.

6.2.1 Method Description

Let us first describe the method employed for the evaluation and break it down according to the presentation by Isdahl & Kjensmo (2019). As mentioned previously, the methodology in Isdahl & Gundersen (2019) builds on the method proposed by Gundersen & Kjensmo (2018), which is why references is made to both of these papers to provide a holistic and complete view of the evaluation method.

To quantify the reproducibility of results presented in a sample of 400 AI research papers, Gundersen & Kjensmo (2018) chose to distinguish between three degrees of reproducibility as explained in the list below.

- **R1: Experiment Reproducible** - This is achieved when an independent researcher is able to generate the same results as the original researcher by using the same implementation of the AI method on the same data. Results that are R1 reproducible require a documentation of everything that is needed to re-run the experiment. When re-running the experiment, the results should be exactly the same as in the original experiment. This degree of reproducibility is most consistent with our definition of reproducibility.

- **R2: Data Reproducible** - This is achieved when an independent researcher is able to generate the same results as the original researcher by using a different implementation of the AI method on the same data. Results that are R2 reproducible require a documentation of the method and data that has been used for the experiment.

- **R3: Method Reproducible** - This is achieved when an independent researcher is able to generate the same results as the original researcher by using a different implementation of the AI method on different data. Results that are R3 reproducible only require a documentation of the method that has been used for the experiment.

The view of Gundersen & Kjensmo (2018) states that R1 reproducible results offer the lowest amount of generality and R3 reproducible results offer the highest amount of generality. In this case, an increased generality indicates that the performance of the AI method is not strictly related to the implementation of it or the data that is used. They also argue that R1 represents the highest amount of transparency and that R3 represents the lowest amount of transparency. This means that R1 reproducible results demand the most detailed documentation to be fulfilled, as it requires a description of everything that is needed to re-run the experiment. In summary, the relationship between these three reproducibility metrics in terms of generality and required documentation can be described by equations (1) and (2), respectively.

$$R1 < R2 < R3 \tag{1}$$

$$doc(R3) \subset doc(R2) \subset doc(R1) \tag{2}$$

Based on our definition of reproducibility as stated in Section 1, the R1 degree of reproducibility is most in line with how we have chosen to distinguish reproducibility from replicability. This is also the degree that demands the highest level of transparency, which is also an aspect that we have underlined throughout this paper as an important component to ensure reproducibility. Gundersen & Kjensmo (2018) also state that R1 corresponds to what Peng (2011) calls full reproducibility with regards to the spectrum of reproducibility, which is another concept that is connected to how we use the term reproducibility.

Furthermore, Gundersen & Kjensmo (2018) identified 16 boolean variables they deemed as important indicators for reproducibility. They grouped these variables under three factors labelled *Experiment*, *Data* and *Method*, which are, in this context, related to a set of documentation activities that are important for reproducibility: *Method* describes the reporting of methods and ideas to other researchers; *Data* refers to sharing of used data and clarifying which parts that were used for training, validation and testing; and *Experiment* refers to sharing of code both for running the experiment and for any methods that are developed. Inspired by Gundersen & Kjensmo (2018) – who used the variables to asses the documentation practices found in research papers – Isdahl & Gundersen (2019) used the same notions of R1, R2 and R3, but modified and extended the reproducibility variables from 16 to 22 so that they could appropriately be used to evaluate how well machine learning platforms support *"reproducible empirical research by scoring the platforms on whether they have features that satisfies the variables"* (Ibid, p. 2). These variables can be seen in Table 4 grouped under the factor that they are related to.

Isdahl & Gundersen (2019) claim that some of the variables in Table 4 are automatically satisfied if the platforms are integrated with specific external systems. By external systems, they refer to notebooks (for example Jupyter), source code management (SCM, for example Git and Github), Docker and TensorBoard. Which variables that are satisfied through the integration of these external systems, according to Isdahl & Gundersen (2019), can be seen in Table 5.

Gundersen & Kjensmo (2018) also proposed six metrics to use for quantifying whether an experiment e is $R1$, $R2$ or $R3$ reproducible. First, they suggested three boolean metrics $R1(e)$, $R2(e)$ and $R3(e)$ that can be either true or false depending on the documentation of the experiment. These are described by equations (3), (4) and (5), respectively; since Isdahl & Gundersen (2019) use the variable p to represent a particular machine learning platform instead of an experiment e, all occurrences of e in the equations have been substituted with p:

$$R1(p) = Method(p) \wedge Data(p) \wedge Exp(p), \tag{3}$$

Table 4: *22 variables identified by Isdahl & Gundersen (2019) as important indicators for reproducibility support found in machine learning platforms, each accompanied with a description as formulated by Isdahl & Gundersen (2019). Each variable is related to one of the factors labelled as Method, Data and Experiment.*

Factor	Variable	Description
Method	Hypothesis	Document the hypotheses to be assessed.
	Prediction	Document the predicted outcome of the experiment.
	Setup	Parameters and the conditions to be tested and desired statistical significance of the results.
	Problem description	Support description of the problem to be solved.
	Outline	Support for outlining the method conceptually.
	Pseudo code	Support for describing the AI method as pseudo code.
Data	Data repository	Share data in a community repository.
	Data metadata	Include basic metadata that describes the data.
	Data license	Give the data a license.
	Data citeable	Generate a digital object identifier (DOI) or persistent URL (PURL) for the version used.
Experiment	Results	Document the results (i.e. measures and metrics) and the analysis.
	Analysis	Explicitly indicate whether the analysis supports the hypotheses.
	Justification	Validate that the chosen datasets, empirical design, and the metrics are appropriate for assessing the results.
	Workflow	Workflow representation that summarizes how the experiment is executed and configured.
	Workflow execution	Workflow execution traces providing settings and initial, intermediate, and final data.
	Hardware	Document the hardware used for running the experiments.
	Software	Document the software dependencies.
	Experiment citeable	Automatically generate reference entry for experiment.
	Code repository	Shared code through community repository.
	Code metadata	Include basic metadata for describing the code.
	Code license	Include a license.
	Code citeable	Generate a digital object identifier (DOI) or persistent URL (PURL) for the version used.

$$R2(p) = Method(p) \wedge Data(p), \tag{4}$$

$$R3(p) = Method(p), \tag{5}$$

where $Method(p)$, $Data(p)$ and $Exp(p)$ is the conjunction of the variables for each factor seen in Table 4. For example, in order for $Data(p)$ reproducibility to be supported for a machine learning platform p, all variables listed in the Data row in Table 4 have to be true. As can be seen from the equations above and which was mentioned earlier, $R1$ is the most strict requirement as it demands that all variables for all factors in Table 4 must be true.

The boolean metrics in equations (3), (4) and (5) only indicate whether a platform p supports $R1$, $R2$ or $R3$ reproducibility in a strict sense. To circumvent this, Isdahl & Gundersen (2019) adopted three final metrics from Gundersen & Kjensmo (2018) to measure the *degree* to which a platform supports $R1$, $R2$ and $R3$ reproducibility. They label these metrics $R1F(p)$, $R2F(p)$ and $R3F(p)$, and these are, in principle and a mathematical sense, identical to the original metrics used by Gundersen & Kjensmo (2018) to measure the degree of reproducibility. The three metrics for measuring the degree of reproducibility are presented by equations (6), (7) and (8), respectively:

$$R1F(p) = \frac{\delta_1 Method(p) + \delta_2 Data(p) + \delta_3 Exp(p)}{\delta_1 + \delta_2 + \delta_3}, \tag{6}$$

Table 5: *A collection of reproducibility variables from Table 4 that are satisfied through the integration of external systems, according to Isdahl & Gundersen (2019).*

System	Variables
Notebooks	Justification, Hypothesis, Prediction, Problem description, Outline and Pseudo Code
SCM	Code repository, Code metadata and Code license
Docker	Software dependencies
TensorBoard	Workflow, Results

$$R2F(p) = \frac{\delta_1 Method(p) + \delta_2 Data(p)}{\delta_1 + \delta_2}, \tag{7}$$

$$R3F(p) = Method(p), \tag{8}$$

where $Method(p)$, $Data(p)$ and $Exp(p)$ are the weighted means of the truth values for the boolean variables found in Table 4, for which the weights are given by δ_1, δ_2 and δ_3. Using these metrics to evaluate our solution, we can score the STACKn platform on every variable seen in Table 4 based on whether STACKn supports features that cover the functionality of each variable. Both Isdahl & Gundersen (2019) and Gundersen & Kjensmo (2018) used uniform weights $\delta_i = 1$ in their respective survey in favor of assigning each variable and factor a different weight. However, note that Isdahl & Gundersen (2019) are clear on this point: A uniform weight might not make sense since some features obviously are more important for reproducibility support. They chose a uniform weight in spite of this observation with the explanation that finding the appropriate set of weights could be a research project on its own, and that was not the intention with their study. We accept the motivation and reasoning behind their weight assignments, and thus, we also use this motivation for using a uniform weight of $\delta_i = 1$ when evaluating STACKn; this is appropriate considering that the purpose of our evaluation is very similar to the one by Isdahl & Gundersen (2019), and investigating which set of weights that is ideal is far out of the scope of our study. We also do this to avoid a biased, and possibly incorrect, decision of which features are more important for reproducibility.

Finally, Isdahl & Gundersen (2019) based their scoring on whether a feature in a specific platform is *Supported*, *Partially supported* or *Not supported*. These should determine the score of the truth values in Table 4, where supported features get a score of 1, partially supported features get a score of 0.5, and not supported features get a score of 0. Partially supported features could either be partially supported as part of the machine learning platform or supported through *active* integration with third-party software. Table 6 and 7 summarize the results of the evaluations Isdahl & Gundersen (2019) conducted on the ML platforms they included in their analysis. These results are included for reference when we evaluate the degree of reproducibility that STACKn supports.

Table 6: *Mean of the variables supported in 13 ML platforms that are similar to STACKn. Values retrieved from Table IV in Isdahl & Gundersen (2019).*

Platform	Experiment	Data	Method
OpenML	0.25	0.75	0.17
MLflow	0.42	0.00	0.58
Polyaxon	0.38	0.00	0.58
StudioML	0.33	0.00	0.58
Kubeflow	0.50	0.00	0.58
CometML	0.67	0.00	0.58
Amazon SM	0.29	0.00	0.58
Google CML	0.33	0.00	0.50
Azure ML	0.42	0.00	0.58
Floydhub	0.71	0.75	0.50
BEAT	0.71	0.75	0.50
Codalab	0.67	0.75	0.50
Kaggle	0.63	0.75	0.50

Table 7: *Degree of reproducibility supported in 13 ML platforms that are similar to STACKn. Values retrieved from Table V in Isdahl & Gundersen (2019).*

Platform	R1F	R2F	R3F
OpenML	0.39	0.46	0.17
MLflow	0.33	0.29	0.58
Polyaxon	0.32	0.29	0.58
StudioML	0.31	0.29	0.58
Kubeflow	0.36	0.29	0.58
CometML	0.42	0.29	0.58
Amazon SM	0.29	0.29	0.58
Google CML	0.28	0.25	0.50
Azure ML	0.33	0.29	0.58
Floydhub	0.65	0.63	0.50
BEAT	0.65	0.63	0.50
Codalab	0.64	0.63	0.50
Kaggle	0.63	0.63	0.50

6.2.2 Method Limitations

The method described above should be an appropriate way for us to measure the degree of reproducibility that STACKn supports, but there are some clear downsides with the method. Here, we present some caveats with the method that we have identified and that are worth taking into account during the evaluation of STACKn.

Objectively assigning scores to the features in STACKn is not a completely simple matter, as this requires some kind of subjective reasoning; for example, some features might be difficult to classify as either supported or partially supported and this requires us to make some kind of subjective decision, and this might skew the results. There is also room for interpretation when it comes to the reproducibility variables in Table 4 with regards to exactly what they constitute. For example, should the variable *Code citeable* only be satisfied when a DOI or PURL is present, or should a Git commit hash for code also be seen as a way of citing code? We make efforts to motivate each decision we take with regards to the scoring and how we interpret the variables.

Also, we argue that only three levels to describe whether a feature is supported (0, 0.5 or 1) is a bit too simplistic and the analysis would probably benefit from using the whole range from 0 to 1. Isdahl & Gundersen (2019) are aware of this as they mention this briefly, but they believe that a greater range would have resulted in a subjective score. While this might be true, we think it is worth mentioning as a clear negative aspect of the evaluation method (something that Isdahl & Gundersen (2019) could have emphasized further in their paper), and it would be good to develop a more robust framework with more dimensions to use in future evaluations of reproducibility support in machine learning platforms. With that said, we use the same scoring range as Isdahl &

Gundersen (2019) to describe the features in STACKn; this allows us to avoid unwanted bias and make (relatively) fair comparisons to the platforms that were included in the survey by Isdahl & Gundersen (2019).

Another caveat of the method as it was used by Isdahl & Gundersen (2019) is that they only collected data through the available documentation for each platform (and further investigations in cases where the documentation was unclear). As such, they did not conduct proper and complete experiments with the platforms, creating a substantial risk for unfair evaluation and scoring of the platforms' various features. It is understandable that they based their evaluation on documentation instead of real experiments with the platforms, considering the time resources that would have been necessary to conduct detailed experiments with as many as 13 complex machine learning platforms, many of which are classified state-of-the-art. Nevertheless, this is another clear downside of the survey presented by Isdahl & Gundersen and this is one instance where we deviate from the method as they followed it. As we only have one platform to evaluate (instead of 13), we conduct model experiments in STACKn, partly to illustrate how our implemented solution works, and partly to motivate how we choose to score each reproducibility feature in STACKn according to the framework by Isdahl & Gundersen (2019).

6.3   Analysis Model

When Gundersen & Kjensmo (2018) presented the three degrees of reproducibility that are described in Section 6.2, they claimed that $R1$ reproducibility corresponds to what Peng (2011) refers to as fully reproducible. For results to be $R1$ reproducible, the documentation should cover everything that is necessary to re-run the experiment, which should also produce the *exact* same results. Isdahl & Gundersen (2019) adopted the notion of $R1$ reproducibility and presented $R1F(p)$ as a metric to measure the degree to which an ML platform p supports $R1$ on an interval ranging from 0 to 1. Connecting this to the spectrum of reproducibility proposed by Peng (2011) seen in Figure 1, one could argue that it is possible to use the $R1F$ metric as a numerical measurement scale axis for the spectrum of reproducibility when evaluating machine learning platforms and their support for conducting reproducible experiments.

However, Peng's (2011) spectrum of reproducibility is not specifically designed to describe reproducibility of machine learning research, which is the focus of this paper. Also, Peng's version of the spectrum of reproducibility states that availability of code is the minimum requirement for reproducibility, and that only textual descriptions of an experiment renders it irreproducible. This contradicts the statement by Isdahl & Gundersen (2019, p. 4), namely that notebooks are *"judged by many as a solution for running reproducible experiments and can reasonably be expected to improve communicating experiments to other researchers to some extent."*

To circumvent the issues with Peng's original spectrum of reproducibility seen in Figure 1, we can update it according to Tatman et al. (2018). Like mentioned in Section 2, they suggest an updated version of the reproducibility spectrum that is more in line with current practices for sharing code and data, and that is more suitable for machine learning experiments. Tatman et al. (2018) suggest three levels of reproducibility indicating what should be shared for each level: 1) Low reproducibility: Finished paper only; 2) Medium reproducibility: Paper, code and data; and 3) High reproducibility: Paper, code, data and environment. These levels of reproducibility is, according to Tatman et al. (2018), equivalent to the three levels included in Peng's original spectrum of reproducibility. Also, considering that Tatman et al. (2018) define reproducibility in the exact same way as Isdahl & Gundersen (2019) define $R1$ reproducibility, a correlation between the updated spectrum and the $R1F$ metric is starting to crystallize.

Based on the observed correlation between the updated spectrum and $R1F$, we propose a simple analysis model in Figure 19 where $R1F$ acts as a scale axis for the reproducibility spectrum. Note that we have used the notion of *Notebooks* in this model to represent the minimum requirement for low reproducibility, as opposed to the notion of *Finished paper* suggested by Tatman et al. (2018). This was based on the assumption that a notebook can be seen as equivalent to a research paper in the context of ML platforms and in terms of the kind of documentation they allow for (they both allow for detailed textual descriptions of conducted experiments).

We argue that this model is justified since the reproducibility levels proposed in the spectrum by Tatman et al. (2018) can be translated to the reasoning behind the $R1F$ metric and its underlying components. For example, due to $R1F$ being dependent on $R2F$ and $R3F$, the integration of notebooks in an ML platform would yield $R1F > 0$ even if no experiment code or data is shared, indicating a "Low reproducibility" as indicated in Figure 19.

This model was created as a necessary tool to use with the purpose to provide an answer to the part research question (iv) seen in Section 1.4 that refers to the spectrum of reproducibility. This model is also relevant in more general terms as the spectrum in its original form lacks the kind of quantifiable metric that can make it useful for more practical analytical purposes.

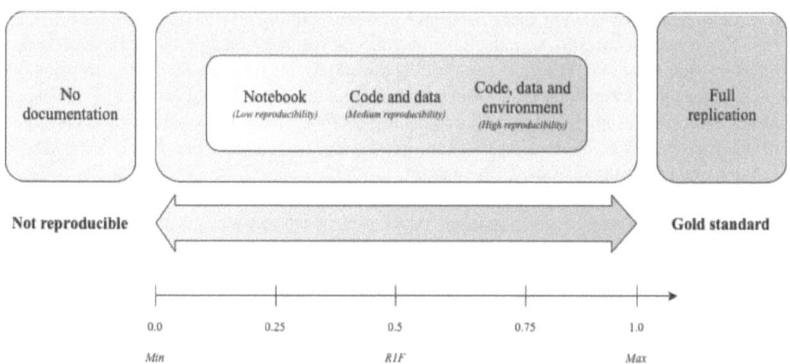

Figure 19: *Analysis model created to evaluate where on the reproducibility spectrum it is possible to place experiments that are conducted in STACKn. The model builds on Peng's (2011) original reproducibility spectrum and the updated reproducibility levels proposed by Tatman et al. (2018), with the addition of the reproducibility degree metric $R1F(p)$ presented by Isdahl & Gundersen (2019). Here, the $R1F(p)$ metric represents a continuous numerical scale axis for the spectrum of reproducibility.*

6.4 Evaluating STACKn

This section is focused primarily on providing an answer to the research question concerning the degree of reproducibility that we can attain after integrating our reproducibility features in STACKn. Using the method presented by Isdahl & Gundersen (2019) as described above, we evaluate four different states of STACKn: First, we evaluate the degree of reproducibility that is supported in STACKn before integrating our features, referred to as the *Initial state*; secondly, we evaluate the degree of reproducibility that is supported in STACKn after integrating only the documentation frameworks Datasheets and Model Cards, referred to as *State 1*; third, we evaluate the degree of reproducibility that is supported in STACKn after integrating only the provenance features that are related to the CLI commands `stackn dvc` and `stackn run`, referred to as *State 2*; finally, we evaluate the degree of reproducibility that is supported in STACKn after a complete integration of the reproducibility features, that is, after integrating the combined solution of documentation frameworks and provenance features, referred to as *State 3*.

Each state of STACKn is evaluated from the perspective of the factors and variables seen in Table 4 in order to calculate a value for the reproducibility metrics $R1F(p_i)$, $R2F(p_i)$ and $R3F(p_i)$, where $i = 0, 1, 2, 3$ each corresponds to a different state of STACKn. All values that are calculated below are rounded to the nearest number to two decimal places. The results of the evaluation indicate whether both of the implemented feature types are necessary for reproducibility and which of these that contribute the most to a high degree of reproducibility. For a complete summary of the evaluation results, see Tables 8, 9 and 10 in Section 6.5.

6.4.1 Benchmark: The Initial State

As a benchmark we wanted to check the degree of reproducibility that the initial state of STACKn supports; in other words, the reproducibility degree that was supported by the application before integrating our solutions. The purpose of testing the initial state for reproducibility was primarily to use the result as a benchmark for the degree of reproducibility we could obtain with our implementation in later testing phases, and draw conclusions for whether our solutions are necessary to include for reproducibility in the application. In this case, $i = 0$ and the platform is referred to as p_0 when calculating the reproducibility metrics.

Method: As mentioned earlier in the paper, the initial state of STACKn uses Jupyter Labs as the primary hub for experimentation in the Projects that the user has created. In the Lab session the user can use Jupyter Notebooks as the primary method to document the experiment workflow through textual descriptions. Because of this feature, some of the variables seen in Table 4 are satisfied in the initial STACKn state. As seen in Table 5, Isdahl & Gundersen (2019) claim that the integration of Notebooks automatically fulfill seven variables, six of them related to the reproducibility factor *Method*; all variables under this factor, except for *Setup*, are satisfied through integration of Notebooks. However, we argue that the variable Setup is fulfilled through the integration of Notebooks as well, considering that the experimental setup with hyperparameters, values, and so on, is easily described in text using a notebook.

As we have indicated multiple times throughout this paper, a Notebook is not a substitute for a structured framework that guides the user in what kind of textual information that should be documented, as it lacks any standardized guidelines for documentation. This is also a sentiment shared by Isdahl & Gundersen (2019) who state that the integration of some kind of Notebook is only able to partially satisfy the *Method* factor, at best. For this reason, we argue that all variables related to *Method* should be labelled as partially supported with regards to the initial STACKn state.

Based on the above motivations regarding the support for the variables in this factor, the reproducibility metric $R3F(p_0)$ is calculated according to equation (9).

$$R3F(p_0) = Method(p_0) = \frac{6 \cdot 0.5}{6} = 0.50 \tag{9}$$

Data: When it comes to the factor *Data* for the initial state of STACKn, there are clear limitations in the variables that are satisfied. A positive aspect is that STACKn offers users the possibility to host data in the S3 compatible object storage MinIO, but this feature does not offer any way to version data and track changes in data, no way to provide structured metadata describing the datasets, no data licenses and no possibility to cite the data. Due to this, we argue that the only variable that is satisfied for this factor is *Data repo* that indicates whether the platform offers a community repo to share data (in this case MinIO). We claim this variables is only partially supported as this kind of cloud storage only enables sharing data within a data science team that are members of the same STACKn Project, and does not support public sharing of data. One could make the case that *Data metadata* is supported to some extent since notebooks let users document and describe how the data was pre-processed before analysis; while this information is important to document and communicate, we argue that this non-exhaustive information alone does not justify scoring the variable as supported, neither partially or fully.

Based on the above motivations regarding the support for the variables in this factor, the reproducibility metric $R2F(p_0)$ is calculated according to equation (10).

$$R2F(p_0) = \frac{\delta_1 Method(p_0) + \delta_2 Data(p_0)}{\delta_1 + \delta_2} = \frac{0.5 + \frac{0.5}{4}}{2} = 0.31 \tag{10}$$

Experiment: To cover the variables under this factor, both textual documentation and other documentation practices are necessary for highly transparent reporting. We claim that the only variables that are partially supported through the integration of Jupyter Notebooks are *Results*, *Analysis* and *Justification*. These variables should be possible for users to document in their notebooks, but once again, there is no standardized approach for what should be documented or how the documentation should be made. However, the variable *Software* is fully supported in the STACKn platform as it utilizes a virtual environment through Jupyter Lab Session as a hosting service where users can collaborate in notebooks and run experiments. As such, every researcher that runs an experiment in STACKn will be using the same software dependencies.

One could argue that the variables *Workflow* and *Workflow execution* are also partially supported through integration of notebooks, but the manual input for this would be too complex and time-consuming to be seen as actively supported element in STACKn. None of the other variables in the factor *Experiment* are satisfied, with the following motivations. Regarding *Hardware*, there is no automatically generated information on the hardware that is used when running an experiment on a local machine; of course, the user could document the hardware specifications manually in theory, but it appears like something few users would actually want to do when running an experiment. However, if the user wants to run experiments on the Kubernetes cluster instead of locally, the user can choose between different hardware specifications (CPU and Memory) to start an instance on the cluster with a virtual environment through a Jupyter Notebook session. This makes it difficult to score this variable, but we choose to label *Hardware* as not supported in the initial state since there is no hardware documentation when running STACKn locally. The variable *Experiment citeable* is also not supported since a reference entry is not automatically created when an experiment is run; as such, there is no way users can refer to previously conducted experiments at a later stage. The variables *Code repository*, *Code metadata*, *Code license* and *Code citeable* are not supported either since there is no integration with any SCM such as Git that provides experiments with pointers to which version of code that was run for the experiment.

Based on the above motivations regarding the support for the variables in this factor, the reproducibility metric $R1F(p_0)$ is calculated according to equation (11).

$$R1F(p_0) = \frac{\delta_1 Method(p_0) + \delta_2 Data(p_0) + \delta_3 Exp(p_0)}{\delta_1 + \delta_2 + \delta_3} = \frac{0.5 + \frac{0.5}{4} + \frac{2.5}{12}}{3} = 0.28 \qquad (11)$$

6.4.2 State 1: Documentation Integration

The second state that was evaluated revolves around the solutions implemented to ease and facilitate documentation of the models trained in STACKn, as well as the datasets used to generate the models. The provenance features built for logging the lineage of machine learning experiments are therefore not included in this evaluation. The two frameworks in focus here are Datasheets and Model Cards, and thus, the results of this evaluation is somewhat difficult to quantify since any gain in reproducibility acquired from these tools depend not on the tools themselves, but on input from the user. To circumvent this issue, we assume that the documentation provided via the Model Card and Datasheet is covering as much information as possible, given how the two forms are composed in the user interface (that is, we do not assume that the documentation contains any information that the user has not been prompted to submit). In this case, $i = 1$ and the platform is referred to as p_1 when calculating the reproducibility metrics.

Method: With the added functionality of writing Model Cards and Datasheets, carried out with supporting question prompts, this reproducibility factor is marginally improved. The variable *Problem description* sees increased support compared to the initial state. This is thanks to the introduction of the Model Card feature, where the user can specify intended use of the model and what problems said model is constructed to solve. Therefore, this variable is set to the value of 1 (increasing from 0.5 in the benchmark). The other variables within the factor; *Hypothesis, Prediction, Setup, Outline,* and *Pseudo code*, remain on their respective level of support present in the initial state.

Based on the above motivations regarding the support for the variables in this factor, the reproducibility metric $R3F(p_1)$ is calculated according to equation (12).

$$R3F(p_1) = Method(p_1) = \frac{5 \cdot 0.5 + 1}{6} = 0.58 \qquad (12)$$

Data: With the introduction of datasheets, this factor is also subject to marginal improvement compared to the benchmark. *Data metadata* can easily be added via the creation of a datasheet. However, in its current form, the datasheet form does not support the generation of a *Data citeable* or a *Data license*. These two variables are for now left as suggestions on further development of the datasheet generation in STACKn. Hence, the two supported variables in the factor *Data* for this case is *Data repo* and *Data metadata*, with *Data repo* still being only partially supported while *Data metadata* enjoys full support.

Based on the above motivations regarding the support for the variables in this factor, the reproducibility metric $R2F(p_1)$ is calculated according to equation (13).

$$R2F(p_1) = \frac{\delta_1 Method(p_1) + \delta_2 Data(p_1)}{\delta_1 + \delta_2} = \frac{0.58 + \frac{1.5}{4}}{2} = 0.48 \tag{13}$$

Experiment: Few of the variables present in this factor are affected by the newly implemented documentation frameworks, and no major increase in the isolated reproducibility metric $Exp(p_1)$ is noted compared to the benchmark. However, we argue that the variable *Justification* can be improved to fully supported, since the Datasheet feature lets users describe and justify why the chosen datasets are appropriate to use for analysis purposes, whether it be training data, validation data or test data. Furthermore, the Model Cards feature prompts users to describe in detail why the chosen metrics were appropriate for assessing a model's performance, something that should be documented under the Metrics section in the model card. As such, we find it justifiable to increase the score for the variable *Justification* after integrating the documentation frameworks, from partially supported to fully supported as compared to the benchmark state. The variable *Results* could have been improved after integrating Model Cards if the section Quantitative analyses had been included in the model cards as suggested by Mitchell et al. (2019). This would have allowed users to include images or plots of interesting metrics that indicate how the model performs. However, this model card feature was excluded due to time constraints.

Based on the above motivations regarding the support for the variables in this factor, the reproducibility metric $R1F(p_1)$ is calculated according to equation (14).

$$R1F(p_1) = \frac{\delta_1 Method(p_1) + \delta_2 Data(p_1) + \delta_3 Exp(p_1)}{\delta_1 + \delta_2 + \delta_3} = \frac{0.58 + \frac{1.5}{4} + \frac{3}{12}}{3} = 0.40 \tag{14}$$

6.4.3 State 2: Provenance Integration

Here, the evaluation revolves around the STACKn state that emerges after integrating the new CLI commands `stackn run` and `stackn dvc` which together constitute the provenance features that are implemented in STACKn. Just like in the previous evaluation where we calculated the effect that documentation frameworks have on the reproducibility in STACKn, we here isolate the effect of the provenance features to see how they alone affect the degree of reproducibility. In this case, $i = 2$ and the platform is referred to as p_2 when calculating the reproducibility metrics.

Method: This factor is not majorly impacted by the provenance features compared to the initial benchmark state, since most of these variables can most easily be satisfied through textual documentation. The case could be made that the variable *Setup* should be improved through the use of the TrackingClient as it allows for logging of which hyperparameters that are used for an experiment. However, this variable demands that more information than only parameter setup is documented for it to be satisfied, making us hesitant to increase the variable's score from partially supported to fully supported. This is clearly an instance where the evaluation would benefit from a scoring interval with greater range as it would have allowed us to increase the variable's score from 0.5 to a score somewhere in the interval [0.5, 1], for example 0.7 to indicate a slight improvement from the initial benchmark state. Besides *Setup*, all the variables under this factor are clearly not affected by STACKn's provenance features, as they all require textual input for proper documentation.

Based on the above motivations regarding the support for the variables in this factor, the reproducibility metric $R3F(p_2)$ is calculated according to equation (15), being the same as $R3F(p_0)$ for the factor *Method* in the initial state.

$$R3F(p_2) = Method(p_2) = \frac{6 \cdot 0.5}{6} = 0.50 \tag{15}$$

Data: Regarding the variables *Data repo* and *Data citeable*, one could argue that it is positively affected by the implemented CLI command `stackn dvc` as it allows the user to capture the versions of data in their Git commits. While DVC does not provide a *dedicated* data repository, it utilizes the functionalities of Git to create lightweight and functionally equivalent substitute for a data repo. Considering that the Git commit hash that refers to the experiment code is automatically generated using the `stackn run` command, users can easily investigate what version of the data that were used for a specific experiment by checking out the corresponding Git commit where .dvc files point towards what data versions were used. This also allows users to cite the exact version of the data by referring to its tag as applied to the data when running `stackn dvc`. With this motivation, we argue that the variables *Data repo* and *Data citeable* are improved through the provenance features in STACKn, specifically that the former is fully supported while the latter is partially supported. The reason for scoring *Data citeable* as only partially supported is because it does not include any DOI or PURL, which should be included for the variable to be fully supported according to Isdahl & Gundersen (2019).

When it comes to the variables *Data metadata* and *Data license*, we argue that they are still not supported. The argument could be made that *Data metadata* is supported to some extent since notebooks allow for textual documentation of how data is pre-processed before using it for model training, and also because the pre-processing scripts as used in the experiments are easily tracked for each experiment through the Git commit hash. However, as stated previously in the evaluation for the initial STACKn state in Section 6.4.1, we argue that the pre-processing information alone is not enough for justifying partial or full support of this particular feature.

Based on the above motivations regarding the support for the variables in this factor, the reproducibility metric $R2F(p_2)$ is calculated according to equation (16).

$$R2F(p_2) = \frac{\delta_1 Method(p_2) + \delta_2 Data(p_2)}{\delta_1 + \delta_2} = \frac{0.50 + \frac{1.5}{4}}{2} = 0.44 \tag{16}$$

Experiment: After integrating provenance features in STACKn, there is a clear improvement in the support for some variables in this factor, while others require more interpretation and subjective input in order to judge. First off, it is clear that the variable *Analysis* is not affected at all by the provenance features we have implemented when compared to the initial state. This variable requires textual input from the user since it requires an explicit mention and interpretation of whether the analysis supports the hypotheses, and STACKn's newly implemented provenance features do not simplify textual documentation. The same motivation can be used for the variable *Justification* which is not improved compared to the initial state.

The variable *Results* on the other hand is more challenging to score after integrating the provenance features. Using the provenance features in STACKn, users can easily log parameters, metrics and trained model objects for each conducted experiment, information that can also be easily revisited in the future. In that sense of the word, STACKn supports this variable. However, there is no active support for logging results from model validation or prediction, although this would be quite easy to implement with our Tracking Client. Despite this, we argue that it should be quite easy to generate the same results from model validation and prediction if the setup and results from model training sessions are documented appropriately, which we claim is the case when the `stackn run` and Tracking Client are used as intended. Therefore, we choose to improve the support for the variable *Results* compared to the initial state, as we believe there is enough ground to justify an improvement in the variable.

When it comes to the variable *Workflow*, there is no difference from the initial STACKn state as the platform still lacks an intuitive or automatic way to represent workflows, whether it be graphically (which is the recommended approach) or in text. The variable *Workflow execution*, on the other hand, is slightly improved through the logging functions. As the logging functions can store experiment settings (such as hyperparameters) for each STACKn experiment, one could argue that the support for tracking the workflow execution is improved. Also, if the `stackn dvc` is used as intended, information about how the data is processed and transformed throughout the experiment accompanies each experiment, which also contributes to a better support for the tracking of workflow executions. However, we do not increase the support for this variable to fully supported, since it is difficult to interpret exactly what Isdahl & Gundersen (2019, p. 3) refer to with *"initial, intermediate and final data"* when they describe this variable. For example, as performance metrics and parameters (such as training loss or model error) are constantly improving or deteriorating during model training as the program works through the defined epochs and batches, one could argue that such changes that occur during model training should be documented automatically as well as part of the workflow execution. With this motivation, *Workflow execution* is only improved to partially supported.

Moving on, we argue that the variable *Hardware* is partially supported since each experiment that is conducted with the `stackn run` command is automatically tagged with information about the hardware of the machine that was used to run the experiment. As such, hardware details are given both when experiments are conducted locally and in the virtual environment that uses resources from the Kubernetes cluster. The reason for why we still do not want to label it as fully supported is because there might be some important hardware specifications that are not logged and that might be good to have details on for some experiments. Since we used the Python library `psutil` and the `platform` module for retrieving hardware specifications, there are limitations to what hardware details that can be recorded. For this reason, we do not label the *Hardware* variable as fully supported. Furthermore, the variable *Software* is unchanged from the initial state.

The next variable to score is *Experiment citeable*, which should be fully supported using the implemented provenance features since each experiment is tagged with a unique identifier, or so called reference entry – either a random ID that is automatically generated by the program or a user-defined ID – that can be used to reference a previously conducted experiment.

Then there are the variables that revolve around experiment code, and all of these are improved compared to the initial STACKn state. Since each experiment that is conducted with the `stackn run` command is tagged with information about which code version of the code that was executed (through a Git commit hash) and which GitHub repository that stores the code, we argue that the variables *Code repo*, *Code metadata* and *Code license* are all fulfilled and fully supported. Of course, this assumes that the user is currently working in a Git repository when running `stackn run`.

The variable *Code citeable* is more difficult to score in this instance. Like previously mentioned in Section 6.2.2, should the code be citeable if each conducted experiment is tagged with a Git commit hash that points to the version of the code that was used to conduct an experiment? This is the case for the provenance features seen in our STACKn solution, which encourages us to at least label the variable as partially supported; since each experiment is also tagged with details on which Git repo that inhabits the code version, we believe that our motivation for this score is strengthened. However, since no other reference tag such as a DOI or PURL (which Isdahl & Gundersen (2019) suggest including for the code versions), this variable is not labelled as fully supported.

Based on the above motivations regarding the support for the variables in this factor, the reproducibility metric $R1F(p_2)$ is calculated according to equation (17).

$$R1F(p_2) = \frac{\delta_1 Method(p_2) + \delta_2 Data(p_2) + \delta_3 Exp(p_2)}{\delta_1 + \delta_2 + \delta_3} = \frac{0.50 + \frac{1.5}{4} + \frac{8.5}{12}}{3} = 0.53 \tag{17}$$

6.4.4 State 3: Complete Integration

In the final state that is to be evaluated, every developed feature for reproducibility has been integrated into STACKn. While the impact of each feature has so far been evaluated in isolation, this is an evaluation of how these features impact the degree of reproducibility when they are combined in the final integration. In this case, $i = 3$ and the platform is referred to as p_3 when calculating the reproducibility metrics.

Method: This factor sees no major impact after combining the documentation frameworks with the provenance features. This is largely due to the fact that both of these integrated tools barely have any effect on the reproducibility when they are integrated in isolation. However, the variable *Problem description* is improved thanks to the transparency effect from the Model Card feature that prompts users to describe what problems their models solve and what they should be used for.

Based on the above motivations regarding the support for the variables in this factor, the reproducibility metric $R3F(p_3)$ for the final solution implemented in STACKn can be calculated according to equation (18).

$$R3F(p_3) = Method(p_3) = \frac{5 \cdot 0.5 + 1}{6} = 0.58 \tag{18}$$

Data: This factor sees great improvement after combining the documentation frameworks with the provenance features, benefiting from positive effects from both tools. This shows how documentation frameworks and provenance features fulfill different purposes and should be combined for the best possible effect on reproducibility, something we claimed at the beginning of this paper in Section 1.3. The provenance features as implemented in STACKn allows for improved support both with regards to *Data repo* and *Data citeable*, while the integration of Datasheets allows for detailed description and presentation of metadata for the used datasets, consequently improving the support for the variable *Data metadata*. The only variable that is completely unchanged compared to the initial state of STACKn is *Data license*, which still has no support at all.

Based on the above motivations regarding the support for the variables in this factor, the reproducibility metric $R2F(p_3)$ for the final solution implemented in STACKn can be calculated according to equation (19).

$$R2F(p_3) = \frac{\delta_1 Method(p_3) + \delta_2 Data(p_3)}{\delta_1 + \delta_2} = \frac{0.58 + \frac{2.5}{4}}{2} = 0.60 \tag{19}$$

Experiment: This factor also sees great improvement after combining the documentation frameworks with the provenance features, benefiting mostly from the effects from the provenance features related to `stackn run`, `stackn dvc` and the Tracking Client. All variables except *Analysis*, *Workflow* and *Software* are improved compared to the initial state after integrating the new features. Notable is that the only variable that is improved from the integration of documentation frameworks is *Justification*, all the others that are improved are affected by the provenance features. This observation is not in line with the statement by Isdahl & Gundersen (2019) who say that the factors *Method* and *Experiment* are most easily satisfied by textual descriptions. Based on the result of our evaluation, the variables in *Method* definitely need to be satisfied through textual input as the provenance features had no effect on these; on the other hand, the documentation frameworks had very little impact on the factor *Experiment*, contradicting the aforementioned statement by Isdahl & Gundersen (2019). Based on the above motivations regarding the support for the variables in this factor, the reproducibility metric $R1F(p_3)$ for the final solution implemented in STACKn can be calculated according to equation (20).

$$R1F(p_3) = \frac{\delta_1 Method(p_3) + \delta_2 Data(p_3) + \delta_3 Exp(p_3)}{\delta_1 + \delta_2 + \delta_3} = \frac{0.58 + \frac{2.5}{4} + \frac{9}{12}}{3} = 0.65 \tag{20}$$

6.5 Summary

Here, we present a summary and overview of the results of the evaluations performed above. First, we present a simple visualization model and example of how the "workflow of reproducibility" might take shape in STACKn, see Figure 20. This workflow model is based on the demonstration shown in Section 6.1 and summarizes the functionalities and benefits of using the features that we have integrated in STACKn for increased reproducibility. Specifically, the figure visualizes how our newly implemented reproducibility and transparency features in STACKn can be used to track the model lineage and how users can fall back to previous states of a model whenever they want. Providing constant access to the exact same material as was used in previous experiments, such as code and data, our reproducibility features can contribute to better provenance tracking and reusability of machine learning models in STACKn. Secondly, the heat map in Table 8 shows a summary of all the variables from Table 4 that are supported in the different states of STACKn that were evaluated in the section above. Thirdly, Table 9 shows the mean of the variables supported in each evaluated state of STACKn. Finally, Table 10 shows the calculated reproducibility metrics for each state of STACKn that indicates how the implemented solutions have contributed to the degree of reproducibility for machine learning experiments run in STACKn. This overview of the evaluation results is included to show how the different states of STACKn compare to each other and how the implemented features have affected the degree of reproducibility compared to the initial state of STACKn in which no reproducibility features are organically integrated.

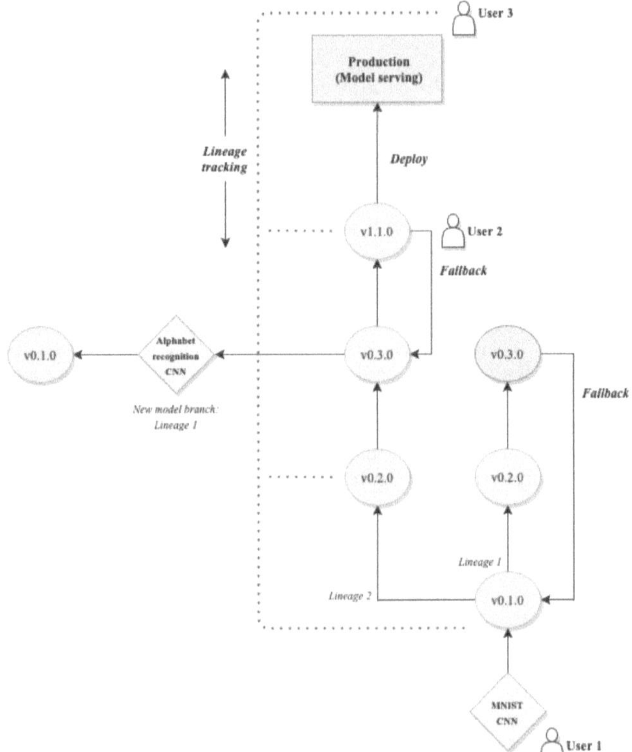

Figure 20: *A simple visualization and example of how the reproducibility workflow might look in STACKn. Assume User 1 creates a MNIST model. The model is trained in Lineage 1 and modified until version v0.3.0 is released. Here, the model version is visualized in red, indicating that the model fails for some reason. Thanks to the reproducibility features in STACKn, User 1 can easily track the lineage of the model evolution and fall back to the initial version of the model. Model training starts again and Lineage 2 starts. The model is trained and eventually put in production when version v1.1.0 is released. User 2, who wants to create an alphabet recognition CNN, tracks the lineage back to version v0.3.0 that has the appropriate characteristics. User 2 retrains the model at this state with new data and creates a model that classifies hand-written letters. Finally, assume User 3 expresses doubts about the model in production. He/she can then track the entire lineage of the model to evaluate the model's validity.*

7 Analysis

This section is dedicated to a detailed analysis of the reproducibility that is supported by STACKn. This section is organized according to the following structure: It begins with a proper analysis of

Table 8: *A heat map indicating what reproducibility features that are either not supported, partially supported or fully supported in the different states of STACKn that have been evaluated. Fully supported features are represented by the darker shade of blue, partially supported features are represented by the lighter blue color, and features that are not supported are represented by white cells.*

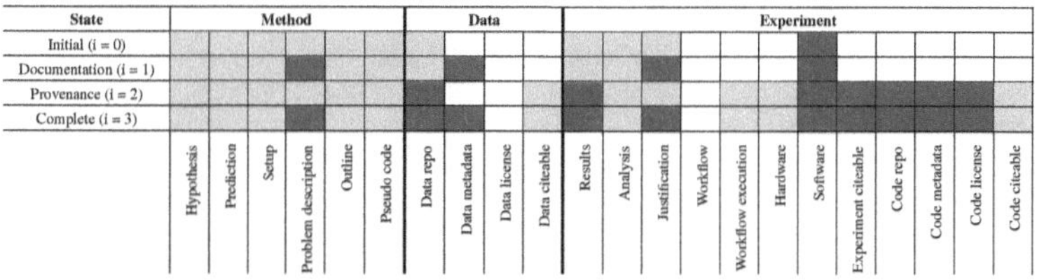

In the table below, ■ = fully supported (dark), ▨ = partially supported (light), and a blank cell = not supported (white).

State	_	Method	_	_	_	_	Data	_	_	_	Experiment	_	_	_	_	_	_	_	_	_	_	_
	Hypothesis	Prediction	Setup	Problem description	Outline	Pseudo code	Data repo	Data metadata	Data license	Data citeable	Results	Analysis	Justification	Workflow	Workflow execution	Hardware	Software	Experiment citeable	Code repo	Code metadata	Code license	Code citeable
Initial (i = 0)	▨	▨	▨	▨	▨	▨	▨				▨	▨	▨				■					
Documentation (i = 1)	▨	▨	▨	■	▨	▨	▨	■			▨	▨	■				■					
Provenance (i = 2)	▨	▨	▨	▨	▨	▨	■			▨	■	▨	▨		▨	▨	■	■	■	■	■	▨
Complete (i = 3)	▨	▨	▨	■	▨	▨	■	■		▨	■	▨	■		▨	▨	■	■	■	■	■	▨

Table 9: *Mean of the variables supported in each evaluated state of STACKn. The highest score for each factor is highlighted in bold.*

i	$Exp(p_i)$	$Data(p_i)$	$Method(p_i)$
0	0.21	0.13	0.50
1	0.25	0.38	**0.58**
2	0.71	0.38	0.50
3	**0.75**	**0.63**	**0.58**

Table 10: *Degree of reproducibility supported in each evaluated state of STACKn. The highest score for each metric is highlighted in bold.*

i	$R1F(p_i)$	$R2F(p_i)$	$R3F(p_i)$
0	0.28	0.31	0.50
1	0.40	0.48	**0.58**
2	0.53	0.44	0.50
3	**0.65**	**0.60**	**0.58**

the evaluation performed in Section 6 and the results thereof. The analysis of the results starts with a general and brief overview of the results, after which it is divided into three parts: 1) First, we analyze and interpret the calculated degree of reproducibility from the perspective of the analysis model presented in Section 6.3; 2) This is followed by a shorter sub-section in which we attempt to compare the reproducibility in STACK with the support for reproducibility found in other existing ML platforms; 3) Finally, we discuss how provenance features and documentation frameworks have affected reproducibility, and whether any of these tools appear to have been the major driving force in improving reproducibility. To end this analysis, we discuss some limitations with the results, which includes a discussion of the validity of our evaluation results and a breakdown of some shortcomings of our proposed analysis model.

7.1 Degree of Reproducibility in STACKn

As the reproducibility degree $R1$ is most in line with how we have chosen to define the notion of reproducibility – that is, that the results should be exactly the same when re-running an experiment with the same material as in the original run – we mostly focus on the reproducibility metric $R1F$ in this analysis. Note also that $R1F$ is dependent on both $R3F$ and $R2F$ and takes all the variables from all three reproducibility factors into account. Hence we argue that this is the most complete and comprehensive metric in Isdahl & Gundersen's framework and the one that mostly corresponds to the updated reproducibility spectrum as proposed by Tatman et al. (2018).

As shown in chapter 6, every newly implemented feature is boosting the reproducibility for at least some variable in some factor of Isdahl & Gundersen's framework. This results in a higher reproducibility degree $R1F$ for each integration that is made, as compared to the initial state of STACKn where no reproducibility features had been integrated. After integrating only documentation frameworks (State 1), $R1F$ increases by 43 %; after integrating only provenance features (State 2), $R1F$ increases by 89 %; finally, after integrating both features for reproducibility (State 3) in STACKn, both provenance and documentation frameworks, $R1F$ increases by as much as 130 % compared to the initial state. Based on these evaluation results, there should be a clear improvement in the ability of machine learning practitioners to reproduce, reuse and validate previously conducted machine learning experiments in STACKn.

At the same time, many variables remain unchanged which goes to show that while the functionality and features implemented as part of this book are a big step towards full reproducibility of experiments and projects run in STACKn, there is still some distance left to be covered before that can be achieved. Through our evaluation, however, what those remaining efforts should consist of is made clear. The most important variables being *Data license* in the factor *Data* as well as *Workflow* in the factor *Experiment* which, even when all implementations are put in place, has no support; a clear indication on what should be focused on if additional development is undertaken to further improve reproducibility in STACKn.

Note that the evaluation results only show a degree of reproducibility that our solutions support in theory. In practice, a lot of responsibility lies in the hands of the user to achieve a high degree of reproducibility, since the users must choose to manually use our newly implemented features in order for some experiment details to be documented. As such, the degree of reproducibility that we report in this section is not necessarily the same degree that other users of the system will experience, but rather what reproducibility features that our solution supports out-of-the-box. Consequently, the evaluation results indicate to which degree that our solution can be used to support users to increase the reproducibility of their experiments, as long as the system is used in the intended way.

7.1.1 The Reproducibility Spectrum

To highlight the results of the evaluation, we can now pinpoint where on the spectrum of reproducibility that it is possible to place experiments that are conducted in STACKn. To do this, we use the analysis model presented in Figure 19 and take advantage of its numerical axis to indicate where on the spectrum STACKn experiments should be placed, basing this on the calculated values for $R1F$ seen in Table 10.

As seen in Figure 21, the reproducibility metric $R1F$ increases steadily as the quantity of integrated reproducibility and transparency features in STACKn increases – from the initial state with no reproducibility features to the final state with complete integration of reproducibility functionalities, there is a clear difference in how reproducibility features promote the support for reproducibility found in STACKn compared to the initial state. What this means is that the features we have implemented in STACKn contributes positively to the degree of reproducibility for machine learning experiments that are conducted in STACKn. Based on the definition of R1 reproducible results presented in Section 6.2.1, this indicates that users of STACKn have a higher chance to generate the same results as seen in previously conducted experiments when using the same implementation of the AI method on the same data.

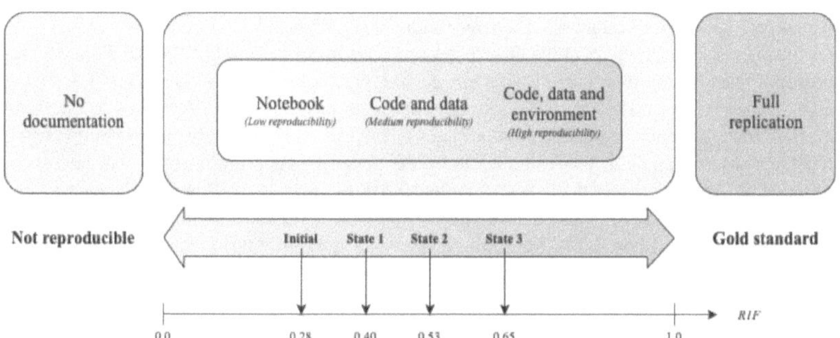

Figure 21: *A visualization of where on the reproducibility spectrum that experiments conducted in STACKn can be placed, indicated by the highlighted points on the axis. Every state of STACKn that was evaluated in Section 6.4 can be seen on the axis, with State 3 supporting the highest degree of reproducibility. State 3 corresponds to the STACKn state in which both provenance features and documentation frameworks have been integrated in the application infrastructure.*

From Figure 21 we see that the positions of each state is well in line with what features that they support in reality. First, look at the initial state of STACKn on the reproducibility spectrum at the far left of the spectrum: It has a fairly low level of reproducibility and seems to be leaning more towards the section of the spectrum where textual documentation in a Notebook is arguably the primary method to reproduce experiments. Now we turn to State 1 where the documentation frameworks Datasheets and Model Cards have been integrated: The ability to add *Problem description* via the Model Card has now been added, along with *Data metadata* through the Datasheet. *Justification* also enjoys increased support. Consequently, this results in a slightly higher level of reproducibility, but it is still on the lower end of the spectrum. Moving on to State 2, with provenance functionality alone implemented. Several variables in the *Experiment* factor is increased, along with a few also in the *Data* factor. These implementations alone pushes the reproducibility level from 0.28 to 0.53 and puts it on the right hand side of the spectrum. With all functionality implemented, the final score of 0.65 is reached, indicated by the right-most arrow in the figure. This is clearly in line with what Tatman et al. (2018) proposes: The availability of code, data and environment is essential for high degree of reproducibility, and this is either partially or fully supported for State 3 of STACKn.

7.1.2 Comparisons with Existing Platforms

Here, we compare the degree of reproducibility calculated in STACKn to the support for reproducibility in other similar ML platforms, primarily to indicate how the support for reproducibility in STACKn stands towards its competitors. Specifically, the other platforms are the 13 that Isdahl & Gundersen (2019) investigated for their survey. We note that STACKn, with its newly implemented reproducibility features, rank on par with the best scoring platforms (BEAT and Floydhub) with $R1F(p_3) = 0.65$. This can be put in relation with the evaluation of the STACKn benchmark test case, with none of our features added, which yielded $R1F(p_0) = 0.28$, for which STACKn would drop to a tie for bottom place together with Google CML.

Just like STACKn, many of the other platforms have integrated Notebooks that automatically create support for the reproducibility factor *Method*, or at least partial support. For this reason, STACKn performs on the same level as many of the other platforms with regards to the factor *Method* seen through a comparison of the values in Tables 6 and 9; no other platform outperforms STACKn when looking at this variable, and only OpenML has a lower score. An interesting observation from Table 6 is that many of the other ML platforms scored low on the factor *Data*. In fact, more than half of them did not satisfy any of the variables that constitute this factor. This is an instance where STACKn, with complete integration, outperforms many of the other ML platforms: In the final version of STACKn, the mean value of the variables that are fulfilled in the factor *Data* is 0.63, compared to many of the other ML platforms that have a value of 0 for this specific reproducibility factor. This could be because the data sharing features have been implemented in a good way in STACKn compared to the other platforms that might not have prioritized such features. However, this could also be a result of Isdahl & Gundersen (2019) interpreting the variables for the factor *Data* in an entirely different way than us.

On the contrary, STACKn scores very low on the variable *Workflow* compared to the other ML platforms. In fact, the evaluation by Isdahl & Gundersen (2019) shows that all investigated platforms except Codalab and Kaggle have integrated support for workflow, which outlines how experiments are executed and configured, although only Azure ML and BEAT have full support for this particular feature. Seven of these platforms have integrated TensorBoard, which is probably a great contributing factor as to why they can support workflow representations. So, in order for STACKn to increase its support for workflow representations it might be a good next step to investigate the possibilities for integrating TensorBoard. On this note, we also argue that this variable is indeed an important variable to support for reproducibility since it provides a good overview of all the experimental steps that are critical for independent researchers to know if they want to reproduce experiments.

7.1.3 Provenance vs. Documentation

Assuming that we have used correct weights for calculating the degree of reproducibility in Section 6.4, we note that the provenance features are the most impactful in terms of providing reproducibility support, while the implemented documentation frameworks has less of an impact. This is not surprising considering that the provenance features affect how code and data are tracked throughout the lineage of a machine learning model, and these components are arguably absolutely essential for reproducibility in machine learning and other data science settings (Brinckman et al., 2019; Chard et al., 2015; Donoho et al., 2009; Eglen et al., 2017; Peng, 2011; Stodden & Miguez, 2014; Stodden et al., 2016; Tatman et al. (2018)). Worth mentioning is that both the documentation frameworks and provenance features improve the *transparency* of machine learning models which can serve to increase reproducibility as well. However, while provenance features provide transparency both to experiments and the actual lineage of models, the documentation frameworks are more limited in what kind of information they can provide with regards to these aspects. Clearly, the key reproducibility features that we have implemented are the ones related to provenance and lineage tracking of experiments and models, such as the Tracking Client and data version control, which might be an explanation to why similar provenance features are much more common than documentation frameworks in other ML platforms. Note that this statement is based mainly on the degree of reproducibility calculated with the evaluation method used in Section 6; as such, this conclusion is only true in the context of this book. In reality, it might be the case that documentation frameworks actually do contribute more to reproducibility than what was captured in our evaluation, although it is highly unlikely in the context of machine learning that these concepts would ever be able to contribute more to experimental reproducibility as opposed to well-integrated provenance features.

When it comes to the documentation frameworks integrated in STACKn it is easy to modify the questions that are used to guide users as to what they should document. For example, one might include a section in the Model Cards where the users are prompted to state explicitly whether the model supports the hypotheses. By doing so, there might be ground enough to justify improving some of the reproducibility variables to fully supported, such as *Analysis* in the factor Experiment. Also, we excluded the section of "Quantitative Analyses" as Mitchell et al. (2019) suggest to include in a Model Card as a visualization tool to show how the model performs graphically, for example by plotting False Positive Rate and False Negative Rate. Of course, this feature in a Model Card could be seen as the most important in terms of reproducibility, as it would be easier to determine if the results of future experiments are in fact reproduced.

Unfortunately, this would still not be a guarantee that users would actually disclose the kind of information that the documentation frameworks urge them to do. In fact, we argue that the greatest issue with including documentation features such as Datasheets and Model Cards in a ML platform such as STACKn is that they do not automatically improve the reproducibility; instead, it is entirely up to the users to decide whether they want to utilize these features. This dependence on user input makes it challenging to evaluate exactly how much documentation frameworks actually contribute to reproducibility in STACKn, both with regards to Model Cards and Datasheets; on the note of Datasheets, Gebru et al. (2018) actually do state that facilitating greater reproducibility is only a *secondary* purpose of Datasheets. To actually make them useful for reproducibility purposes, what questions that should be included in the Datasheets and Model Cards in STACKn is something that probably needs a lot of consideration in the future. We did not spend too much time analyzing what to include and what to exclude, but we want to make it clear that this is important take into consideration to increase the chances that users actually use these features for transparent documentation.

Continuing on the subject of documentation frameworks, the case could also be made that they actually make the process of conducting machine learning experiments and ensuring the transparency of models more demanding if they are used in the intended way. Considering that we aimed to implement features in STACKn that, not only enabled reproducibility and transparency, but also simplified these aspects, the documentation frameworks might actually result in the opposite effect that we wanted to achieve. For example, few users will probably feel the urge to fill in an entire Datasheet in full detail for each version of the datasets that are used for machine learning experiments. Since practitioners will probably want to make changes to the datasets continuously, whether it be through pre-processing methods or by adding or removing data, it might not be viable to keep writing new or updating old datasheets to reflect the many changes made to the datasets. This might be a problem, especially if we take the following statements by Isdahl & Gundersen (2019, p. 9) into concern: *"It is important that the platforms do not get in the way of development machine learning systems, as the main reason professionals use them in the first place is to increase the efficiency of the machine learning p ipeline. The platform should support an improve the development of machine learning systems. The value of machine learning plat-forms should not only be to document the experiment for colleagues and independent researchers in order to enable reproducibility. It is important that the value of using the systems outweigh the cost of setting them up for the end user."* To simplify the use of the documentation frameworks and avoid them becoming an obstacle, one recommendation could be that users limit the amount of information that they supply to a Dataseheet; they might, for example, only provide details on how pre-processing was made and from where the data was retrieved. Furthermore, another simi-lar recommendation could be to only generate exhaustive Model Cards for major model versions, as small changes in a model probably do not require any changes in a Model Card.

However, we do not insinuate that Datasheets and Model Cards are redundant, just that one should be careful with how one interprets their impact on the degree of reproducibility in STACKn as seen in the evaluation. They clearly fulfill certain needs in terms of transparency. For instance, using Datasheets to communicate characteristics of datasets that are used to train a model, other users can get an idea of whether there exist bias in the datasets that could have corrupted the trained model's behaviour. Similarly, using Model Cards to communicate characteristics of trained models, other users can get an idea of what a model can be used for and its limitations, which reduces the risk of a model being used in situations or settings where it is not appropriate, hence reducing the risk of machine learning controversies that have become quite common in recent years (see Section 1.2). Also, this can help independent users who are searching for models with a specific use case find what they are looking for: If they simply revisit earlier versions of a model in the Studio UI of STACKn they might come across the Model Card of an old model version that describes a model that possess the exact characteristics the user requests. The user can then quickly use this old model version "as is" by calling the deployment endpoint, or fallback to this specific version of the model to modify and train it according to own needs. This will save both time and money for organizations and smaller data science teams as it promotes the reusability of old models, enhancing productivity and collaboration in machine learning projects.

Even though documentation frameworks fulfill certain needs and do slightly improve the re-producibility as seen in our evaluation, we argue that the preferred way to increase support for reproducibility in ML platforms is through integration of provenance features. This is not surpris-ing as these features as implemented in STACKn meet many of the fundamental best practices for reproducibility presented in Section 2, most importantly the sharing of experiment code and data. The provenance features automate experiments and logging of experiment data, making the process of ensuring reproducibility and the entire experiment process more streamlined and man-ageable. Looking at the provenance features integrated in STACKn, they should allow users to conduct experiments while storing most, if not every, important detail that is needed to revisit the experiment in the future and reproduce the results. By storing all hyperparameters that are essen-tial to generating a model, metrics that show how the model performs, which type the model is, what kind of pre-processing that was made to the training data, and even storing the model object containing information such as weights, independent users can easily go back in the experiment logs and see when and how a model was produced. This removes the need for manual docu-mentation of such data that can quickly become overwhelming, both with regards to quantity and complexity. No human person will eventually be able to manually document millions of param-eters that might be involved when working with complex machine learning models. Luckily, the provenance features in STACKn should from now on manage this kind of data instead, while also keeping every bit of information collected in one place in the Studio database, visible for the users in the Studio UI. So, even though the documentation frameworks showed to improve the degree of reproducibility in STACKn – specifically the variables *Problem description*, *Data metadata* and *Justification* – we make a strong case for prioritizing provenance features in ML platforms, as they undoubtedly and from an objective standpoint actually do contribute to reproducibility. Also, the reproducibility variables that are improved by the documentation frameworks could probably also be improved if available Jupyter notebooks are utilized to the fullest in order to document the experiments.

With all things considered, while the provenance features seem to have a greater and more indisputable effect on the quantifiable reproducibility support, we still argue that both the provenance features and the documentation frameworks contribute to the reproducibility of machine learning experiments in STACKn when used in combination. The more information we can provide about how a model was generated, the easier it can be reproduced and validated; as such, documentation frameworks can certainly be used as a complementary tool for reproducibility in cases when the provenance features are not sufficient to meet all conditions for reproducibility. So, even though the quantifiable effect from the provenance features as calculated in this paper is more certain, we believe that the documentation frameworks contribute to transparency and reproducibility in ways that cannot be quantified with the method we used for the evaluation in this paper. Simply put, we believe that both of these features help optimize the workflow that was visualized earlier in Figure 20.

However, we want to emphasize that the provenance and documentation features that have been implemented in STACKn are still only a first draft of what might be, as this project has only been a first initiative for increasing the support for reproducibility in STACKn. As such, STACKn has certainly not reached its full potential when it comes to the reproducibility support.

7.2 Limitations

Here, we present some identified limitations that will affect the conclusions we can draw from the results. These limitations include both aspects that have negatively impacted the validity of the evaluation results, and also some important criticisms of our analysis model that we want to put forth.

7.2.1 Validity of Results

We believe that the calculated degree of reproducibility reported in Section 6 is somewhat obscured, much due to the limitations of the evaluation method used. Many of these caveats were discussed in Section 6.2.2 but some of them are worth reiterating here now that the quantifiable evaluation has been completed. We believe that there are two of the identified method caveats that are the primarily responsible. One aspect that sows doubt about the results is the fact that only three scores were used in the evaluation of the reproducibility variables: 0 for no support, 0.5 for partial support, and 1.0 for full support. Clearly, this can skew the results to both directions: There were multiple instances where we considered it to be justified to increase score from 0.5, but could not justify to increase it all the way to 1.0. In such a case, the result would be skewed to a lower score. There were also multiple instances where it was the other way around, for which the results would be skewed to a higher reproducibility score. It is obvious that a greater range is needed when scoring the reproducibility variables that Isdahl & Gundersen (2019) identified as important indicators for reproducibility in ML platforms. However, as previously mentioned we did not want to use a greater range in order to avoid bias, as a greater range of scores for the evaluation would have complicated the scoring process significantly, and for many variables the scores would have been bordering on arbitrary.

Another aspect of the method seen as having a major negative impact on the validity of the results is the fact that it constantly demanded subjective input in order to be able to determine on which level a feature is supported in STACKn. While we chose to use this method primarily with the purpose to reduce unwanted bias in the evaluation to the largest extent possible, it was impossible to remove it completely. As the evaluation was carried out by the creators of the reproducibility tools in focus, it was therefore subject to unintentional, albeit unwanted, biases. Efforts were made to avoid this to the highest degree possible, as discussed in 6.2.2. Since many of the variables in the analysis model are judged subjectively, however, complete fairness and objectivity might be unobtainable.

Moreover, the weights that were used to evaluate the degree of reproducibility in STACKn can be disputed. We do not believe that the set of weights used to calculate $R1F$, $R2F$ and $R3F$ was ideal, as there clearly are some variables that are more important for reproducibility than others. For example, based on our analysis model in Figure 19, the variables related to code, data and environment are very important for reproducibility, which is why these should have been given higher weights in the calculations. In machine learning, these components are essential to share to ensure a high reproducibility. Compare this with the variable *Outline* which arguably is not as important to reproducibility, yet it has had an equal impact on the degree of reproducibility in our calculations.

Also, the results achieved during our evaluation of STACKn should be compared with caution to those of the platforms covered by Isdahl & Gundersen's study. As such, these comparisons cannot be used to draw any strong conclusions, although they do serve to highlight how the support for reproducibility in STACKn, roughly, stands towards it's competitors. While we used the same basic idea as Isdahl & Gundersen (2019) to evaluate the support for reproducibility in STACKn, there are some major differences. Isdahl & Gundersen evaluates the different platforms based solely on the documentation available, while the evaluation made in this book is more thorough. While a comparison between the STACKn reproducibility and the 13 platforms investigated by Isdahl & Gundersen (2019) can give an indication of how STACKn stands to other open-source, no strong conclusions should be drawn from such comparisons. The different results from the STACKn evaluations should be put in contrast to each other, and not with the results from other studies.

Furthermore, and perhaps more important, is that we do not know whether the inherent characteristics of the ML platforms are similar enough to STACKn such that it warrants a fair and justifiable comparison. Of all the platforms that were investigated by Isdahl & Gundersen (2019), the only platforms that we know are similar to STACKn are MLflow and Kubeflow, which we described in Section 4. We know how these two platforms work and what concepts they are founded on, giving us more confidence that they can be fairly compared to STACKn. The other platforms might also be similar enough to justify comparisons with STACKn, but we feel it necessary to at least point out the possibility that they might be too different to render the comparisons reliable. However, worth noting is that Isdahl & Gundersen (2019) made the judgement that all the platforms were fair to compare; if we trust their judgement, it should be justifiable to compare STACKn with the other existing platforms.

Once again, the subjective nature of the evaluation method becomes apparent here as well. We do not know exactly how Isdahl & Gundersen (2019) interpreted the reproducibility variables throughout their evaluation and what criteria they used to score the support for the different variables. For example, what criteria had to be fulfilled in order for a feature to be fully supported and not partially supported? Did they judge the variables strictly? Since we do not know this, there is a risk that they judged the variables very differently as opposed to us. If that is the case, the comparison of our results would not be trustworthy. For an ideal situation for comparison with other platforms it would have been necessary to evaluate the other platforms ourselves.

As the inherent characteristics of the evaluation method seem to be major contributing factors to the reduced validity of our results, we here suggest a measure for evaluating the validity of the method by Isdahl & Gundersen (2019). This is also on the theme of reproducibility, as we feel it necessary to investigate how reproducible the method is in itself. For instance, one could create a survey where a selection of independent individuals are prompted to use the method to evaluate the reproducibility in STACKn (or any other ML platform) to see what results each one reach. If there is a large majority who reach different results in the evaluation, there are some major issues with the method (assuming the results of the survey is statistically significant). We believe that this would probably be the case, unless the method is developed and enhanced further. For example, the method probably needs a lot more structure and strict guidelines for scoring variables in order to remove as much subjective input from the user as possible.

7.2.2 Criticism of Analysis Model

We have identified a number of flaws in the analysis model that is used in Figure 21 and in how it has been applied to analyze the reproducibility support in STACKn, which is why we dedicate this section to discuss these shortcomings. Clearly, the analysis model included is not perfect and it became increasingly apparent throughout evaluation that it should be used with caution for actual analytical purposes.

First off, while we believe that it was a good decision to use the updated version of the spectrum of reproducibility according to Tatman et al. (2018) instead of the original version, we have realized that the neither versions of the spectrum might be such a good benchmark for evaluating the degree of reproducibility found in ML platforms. According to Peng's (2011) original spectrum of reproducibility, it should be possible to achieve full reproducibility as long as a publication is accompanied by linked and executable code and data. Similarly, the reproducibility levels proposed for the reproducibility spectrum by Tatman et al. (2018) state that paper, code, data and execution environment should be shared for a high reproducibility; while we substituted "paper" as used by Tatman et al. for "notebook" in our analysis model to suit the evaluation context of this book paper better, the principle is the same. Compare this with the reproducibility variables in table 4 identified by Isdahl & Gundersen (2019): Arguably, it is not possible to satisfy all of these variables only through the use of a notebook, code, data and the execution environment. For example, the variable *Experiment citeable*, that is fulfilled if experiments are automatically tagged with a reference entry, cannot be satisfied through the components seen in the spectrum model. While Isdahl & Gundersen (2019) did not include any detailed explanation as to why their proposed variables are good for reproducibility (something that would have been desirable for some of the variables), we believe that their exhaustive list of reproducibility variables better represents all the conditions that need to be fulfilled for high reproducibility. As such, the spectrum of reproducibility might not be such a good instrument to indicate the degree of reproducibility for an experiment; we argue that it provides a too simplistic guideline for what to document and share in order to ensure future reproducibility of experiments and the results thereof.

Also, assume an ML platform fulfills all reproducibility variables except the ones related to environment. If we use the $R1F$ metric as a scale for the spectrum, it still indicates that the platform supports high reproducibility and that the execution environment has been shared, when this is not the case. In other words, using the spectrum in combination with the $R1F$ metric as scale can easily create misleading results. The reproducibility spectrum might actually be a more appropriate tool if it is only used for theoretical purposes to visualize the concept of a reproducibility variable that can be measured as a continuous variable, and not analytical purposes. Instead, the $R1F$ metric should probably be used in isolation for analysis and it could even be seen as a spectrum of reproducibility in itself.

8 Conclusions

At the outset of this project, the purpose was to evaluate the possibility of introducing features for reproducibility in STACKn – a cloud-native application for machine learning that runs in Kubernetes – and architect and implement a technical solution in the system that can work together seamlessly with the existing platform architecture. Our ambition was to introduce reproducibility features and technical characteristics in STACKn that are commonly seen in existing provenance systems for machine learning, as well as a built-in feature that facilitates and standardizes the process of documenting the end-to-end ML workflow when using STACKn for ML models. We aimed to present a final solution in STACKn that works well and could actually be of value for the stakeholders of STACKn with regards to requirements for reproducibility and transparency.

To reach our goal of implementing a solution for provenance, transparency, and reusability in STACKn, we defined a number of research questions that we sought to find answers to. We reiterate these questions in the list below.

(i) How can we integrate provenance features typically seen in existing ML platforms, such that ML experiments can be easily reproduced by tracking the lineage of models that are created in STACKn?

(ii) How can we integrate an API for easy manual documentation in STACKn that encourages users to conduct transparent textual reporting of model characteristics, while increasing reproducibility in the process?

(iii) To what *degree* is it possible to achieve support for reproducible empirical research in STACKn, assuming adherence to the definitions of reproducibility as stated in Section 1.1? More specifically, where on the spectrum of reproducibility as presented in Figure 1 can we place experiments conducted in STACKn using the implemented solutions? Is it necessary to apply both provenance tracking features and manual documentation to obtain a high degree of reproducibility? How does it compare to reproducibility support seen in other similar machine learning platforms?

Adhering to the definition of reproducibility as the ability of independent researchers to re-run previously conducted experiments with the exact same material as the original researcher, we set out to answer the research questions above. To do this, we began by looking for inspiration in existing tools used for reproducibility and transparency in machine learning settings. We then conducted an exhaustive software engineering process where we defined the requirements and the architectural design, which were used as a starting point for the software development phase where we implemented technical features for reproducibility in STACKn. Finally, the implemented solution was evaluated using a method suggested by Isdahl & Gundersen (2019) with which the degree of reproducibility could be quantified. Using the so-called spectrum of reproducibility as foundation, we also proposed a model for analysis that visualizes the degree of reproducibility that a ML platform supports.

Measuring reproducibility in an exact manner is a, to say the least, difficult task, and so it remains. Even the term itself is not consistently defined in the related literature. However, a number of measures that can improve reproducibility were identified from existing literature and reproducibility guidelines, and subsequently implemented in STACKn. The evaluation showed that the support for creating reproducible machine learning experiments in STACKn increased as a result. Unfortunately, the validity of the results are not immune to scrutiny, much due to a number of identified limitations in the method used for evaluation. However, while the exact increase and level of reproducibility turned out to be challenging to quantify objectively in STACKn, we argue that there is an indisputable increase in the support for reproducibility of experiments in STACKn thanks to our contributions: Even if the quantifiable evaluation results can be disputed and criticised, we have illustrated how the reproducibility features in STACKn, if used as intended, can be used to promote model provenance, transparency and reusability, which are all important components for experimental reproducibility in machine learning.

Examining the research questions stated above one at a time, we can conclude that we have found answers to all of the research questions stated above. The answer to each question is presented in a concise manner according to the list below.

(i) With inspiration taken from similar systems, where provenance tracking is in place, the implemented Tracking Client has been integrated in such a way that tracking and logging artifacts that are generated in model experiments becomes a simple and manageable undertaking. Lineage artifacts include hyperparameters, parameters, metrics, and model objects. Tracking of the model's performance over time has been made available through plots that visualize how model metrics vary over time. In addition, automatic logging of other important details such as code versions and hardware specifications simplify further in the process of exactly pinpointing the experiment setup that resulted in a specific model.

(ii) Functionality has been put in place to facilitate for manual documentation, both in the form of Model Cards as well as Datasheets. While the difficulty of these features lies in encouraging the users to undertake these sometimes tedious tasks, as it is in no way compulsory to fill out neither of the two forms. If the user chooses not to fill out the forms, the level of transparency and reproducibility is not affected. Nonetheless, this solution can be seen as more encouraging and helpful to users as opposed to Jupyter notebooks that do not offer any standardized framework for what should be documented or how documentation should be performed.

(iii) Arguably the most interesting research question in the context of this paper, the answer to this question could be provided by calculating the degree of reproducibility according to the method presented by Isdahl & Gundersen (2019). As seen in Table 10 and Figure 21, the reproducibility metric $R1F$ – which is consistent with our definition of reproducibility – increases from 0.28, without any of our implemented features, to a value of 0.65 with all features added. As this metric ranges from 0 to 1, this value corresponds the most to *Medium reproducibility* according to the reproducibility levels in the modified spectrum as proposed by Tatman et. al (2018). After integrating nothing but documentation frameworks in STACKn, a slight increase in reproducibility was noted, from 0.28 to 0.40. The increase was greater when nothing but provenance features were integrated in STACKn, from 0.28 to 0.53. With all features put in place, a comparison with the ML frameworks examined by Isdahl & Gundersen (2019) puts our solution on par with the best performing frameworks from that article. STACKn also outperforms most of the existing platforms with regards to the reproducibility variables under the factor *Data*; on the other hand, most platforms have support for the variable *Workflow*, something that STACKn lacks. An objective and fair comparison with the reproducibility support found in other platforms is probably not possible, however, and the contrast should be seen only as a rough indication on where support for reproducible experiments in STACKn stands among its competitors.

In summary, we implemented both provenance features and documentation frameworks in an effort to increase support for reproducible experiments in STACKn. Our evaluation shows that both of these tools contribute positively to the degree of reproducibility, although we achieved a more indisputable impact from the provenance features. The provenance features fulfill more of the recommendations and best practices proposed in literature and by other stakeholders, while there is more uncertainty in whether documentation frameworks contribute to reproducibility *per se* or primarily to transparency of models and data. However, we do not insinuate that Datasheets and Model Cards are redundant; we do believe that they are a great addition to STACKn, especially as they are easily available in the STACKn Studio UI where they can be used for proper and transparent reporting of datasets and models. As such, we argue that these features do contribute to increasing the transparency of the documentation and, by extension, the explainability of machine learning models in the sense that such documentation practices increase understanding of models and how they have been generated. Simply put, we argue that a combined solution of provenance features and documentation frameworks yields a machine learning environment that promotes model provenance, transparency and reusability.

8.1 Contributions

Through the completion of this book paper, we have contributed with improvements to the STACKn infrastructure, mostly through integration of features that support reproducible machine learning experiments. This should help all users of STACKn to increase the reproducibility of their experiments, no matter the reasons for why they want to promote reproducibility; whether it be researchers in academia who depend on transparent reporting for their scientific findings to be accepted, or company users who want to improve the productivity and efficiency of their data science projects by increasing the reusability and explainability of machine learning models. Furthermore, as stated at the early phase of this paper, we believed that a combined approach of documentation frameworks and provenance features had not been used in any other ML platform so far. With this paper, we have shown that this combination contributes positively to reproducibility, although provenance features had a greater impact on the evaluation results. Nevertheless, we have contributed with the insight that provenance features probably are not enough to ensure high reproducibility, a standardized framework for textual documentation is also needed. And while we also contributed with a modified spectrum of reproducibility to create an analysis model, we argue that it is in great need of further refinement if it should be used in the future for analysis purposes.

8.2 Future Work

Going forward, there are still more functionality that can be implemented in STACKn with regards to reproducibility. Like mentioned previously, STACKn could benefit from integrating support for experiment workflow representation, perhaps through the integration of TensorBoard as seen in many other ML platforms. Also, investigations should be made into how the variables found in the factor *Method* can be increased. These variables can be covered through textual documentation, but our evaluation showed that the documentation frameworks we implemented in STACKn did not do much for these variables. The question emerges on whether it is possible to increase these variables; as we have already mentioned, the textual documentation is left to the user to manage, making it difficult for the system to automatically manage any text-related reproducibility features. To make it as user-friendly as possible, one alternative could be to investigate how one could implement a documentation framework that is integrated with the notebooks that can be accessed used through the Jupyter lab sessions in STACKn. This would make the documentation more streamlined and easily accessible since both code and text documentation would be gathered in one place. As it is now, both Datasheets and Model Cards are disconnected from both the code and the rest of the documentation in the user's notebooks, creating a risk that the user never takes advantage of these features, or might even forget that they exist. There is also a relevance in integrating the functionalities of the provenance features and the documentation frameworks, for example by automatically populating certain sections of a Model Card when the user calls `stackn run`.

On the subject of making things user-friendly, a really good feature to implement in the future would be to create a CLI command which the user can call to automatically run previously conducted experiments. This command could be called something like `stackn reproduce -i <Experiment ID>` for which the user passes the ID of the experiment that should be run, and then the application runs this experiment exactly the same way as was done the first time. This includes transforming data, training a model with this data and with the same setup (hyperparameters, net structure, and so on), validate the model with the same validation data, and then test the model with the same data. As such, the program would need to automatically access the same code and data that was used the first time the experiment was conducted. This can be done using today using the provenance features we implemented as all information is available, but it requires the user to manually trace and extract everything that is needed to rerun an experiment.

Another important thing to improve and make more compatible with STACKn in the future is the `stackn dvc` command. We believe that the DVC system is a really good feature to have, especially when using large datasets in an experiment, so it is certainly worth integrating further in STACKn. As it is today, it only initializes DVC and sets up a correct configuration for the user's current Git and STACKn project; in other words, the DVC library is barely integrated in the STACKn infrastructure, forcing the user to use CLI commands and operations that are native to the DVC library. Today's implementation of DVC is very simplistic, and a good feature to develop in the future is a feature that lets STACKn automatically track the data versions, so that the users do not need to perform DVC native operations such as `dvc add <Data file>` or `dvc commit -m <Commit message>` themselves. It would be good if STACKn managed those kinds of operations on the users behalf. This is also important so that users will not need to learn the DVC native commands.

Aside from further functionalities that could be added to STACKn in the future, there are some other considerations as well. One important example is to conduct user tests by letting users, such as machine learning practitioners or researchers, test the reproducibility features that we have implemented. To not conduct user tests for this book is a flaw that is definitely worth mentioning since a user-friendly design and functionalities that are easy to use are essential for a platform such as STACKn. However, we do not think user tests were directly relevant for the purpose of this paper since we were more focused on the underlying technical functionalities of introducing reproducibility, which we believe justifies the lack of user tests for this particular software project. The usability of the application was a secondary goal, something that we mentioned earlier in Section 1.3.

References

All European Academies (ALLEA) 2017, *The European Code of Conduct for Research Integrity*, Berlin: ALLEA, Available at: https://ec.europa.eu/info/funding-tenders/opportunities/docs/2021-2027/horizon/guidance/european-code-of-conduct-for-research-integrity_horizon_en.pdf (Accessed: 2020-12-29)

Alsheikh-Ali, A.A.; Qureshi, W.; Al-Mallah, M.H. & Ioannidis, J.P.A. 2011, "Public Availability of Published Research Data in High-Impact Journals", *PloS one*, vol. 6, no. 9, pp. e24357-e24357.

Angwin, J.; Larson, J.; Mattu, S. & Kirchner, L. 2016, *Machine Bias*, ProPublica, Available at: https://www.propublica.org/article/machine-bias-risk-assessments-in-criminal-sentencing (Accessed: 2020-12-23)

Ariza, M.; Arroyo, J.; Caparrini, A. & Segovia, M. 2020, "Explainability of a Machine Learning Granting Scoring Model in Peer-to-Peer Lending", *IEEE access*, vol. 8, pp. 64873-64890.

Azodi, C.B.; Tang, J. & Shiu, S. 2020, "Opening the Black Box: Interpretable Machine Learning for Geneticists", *Trends in genetics*, vol. 36, no. 6, pp. 442-455.

Barredo-Arrieta, A.; Lana, I. & Del Ser, J. 2019, "What Lies Beneath: A Note on the Explainability of Black-box Machine Learning Models for Road Traffic Forecasting", *IEEE*, p. 2232.

Beam, A.L.; Manrai, A.K. & Ghassemi, M. 2020, "Challenges to the Reproducibility of Machine Learning Models in Health Care", *JAMA: the journal of the American Medical Association*, vol. 323, no. 4, pp. 305.

Berman, F.; Rutenbar, R.; Christensen, H.; Davidson, S.; Estrin, D.; Franklin, M.; Hailpern, B.; Martonosi, M.; Raghavan, P.; Stodden, V. & Szalay, A. 2016, *Realizing the Potential of Data Science*, Available at: https://www.nsf.gov/cise/ac-data-science-report/CISEACDataScienceReport1.19.17.pdf (Accessed: 2020-12-29)

Braun, M.L. & Ong, C.S. 2014, Open science in machine learning, *Implementing Reproducible Research*, p. 343.

Brinckman, A.; Chard, K.; Gaffney, N.; Hategan, M.; Jones, M.B.; Kowalik, K.; Kulasekaran, S.; Ludäscher, B.; Mecum, B.D.; Nabrzyski, J.; Stodden, V.; Taylor, I.J.; Turk, M.J. & Turner, K. 2019, "Computing environments for reproducibility: Capturing the "Whole Tale"", *Future generation computer systems*, vol. 94, no. C, pp. 854-867.

Chard, K.; Pruyne, J.; Blaiszik, B.; Ananthakrishnan, R.; Tuecke, S. & Foster, I. 2015, "Globus Data Publication as a Service: Lowering Barriers to Reproducible Science", *11th International Conference on e-Science (IEEE 2015)*, August 31-September 4, Munich, Germany, pp. 401-410.

Collins, F.S. & Tabak, L.A. 2014, "NIH plans to enhance reproducibility", *Nature (London)*, vol. 505, no. 7485, pp. 612-613.

Committee on Science, Engineering, and Public Policy (U.S.). Panel on Scientific Responsibility and the Conduct of Research, 1992, *Responsible science: ensuring the integrity of the research process*, National Academy Press, Washington, D.C.

Cush, A. 2016, *This Program That Judges Use to Predict Future Crimes Seems Racist as Hell*, Gawker, Available at: https://gawker.com/this-program-that-judges-use-to-predict-future-crimes-s-1778151070 (Accessed: 2020-12-23)

Donoho, D.L.; Maleki, A.; Rahman, I.U.; Shahram, M. & Stodden, V. 2009, "Reproducible Research in

Computational Harmonic Analysis", *Computing in science & engineering*, vol. 11, no. 1, pp. 8-18.

Driggers, R.G. 2011, "The Importance of Science and Technology", *Optical Engineering*, vol. 50, no. 9, p. 090101-1.

Drummond, C. 2009, "Replicability is not Reproducibility: Nor is it Good Science", *Proceedings of the Evaluation Methods for Machine Learning Workshop at the 26th ICML*, Montreal, Canada

Eglen, S.J.; Marwick, B.; Halchenko, Y.O.; Hanke, M.; Sufi, S.; Gleeson, P.; Silver, R.A.; Davison, A.P.; Lanyon, L.; Abrams, M.; Wachtler, T.; Willshaw, D.J.; Pouzat, C. & Poline, J. 2017, "Toward standard practices for sharing computer code and programs in neuroscience", *Nature neuroscience*, vol. 20, no. 6, pp. 770-773.

Fang, H. & Miao, H. 2020, *Introducing the Model Card Toolkit for Easier Model Transparency Reporting*, Google AI Blog, Available at: https://ai.googleblog.com/2020/07/introducing-model-card-toolkit-for.html (Accessed: 2020-12-19)

Fenster, C.B. & Gropp, R.E. 2020, "On the Importance of Science to Society: A Call for Government Action", *Bioscience*, vol. 70, no. 5, p. 371.

Freire, J.; Fuhr, N. & Rauber, A. 2016, "Reproducibility of data-oriented experiments in e-Science (Dagstuhl Seminar 16041)", *Dagstuhl Reports*, vol. 6, no. 1, pp. 108–159

Gebru, T.; Morgenstern, J.; Vecchione, B.; Wortman Vaughan, J.; Wallach, H.; Daumé III, H. & Crawford, K. 2018, "Datasheets for Datasets", *Proceedings of the 5th Workshop on Fairness, Accountability, and Transparency in Machine Learning*, Stockholm, Sweden.

Goodman, S.N.; Fanelli, D. & Ioannidis, J.P.A. 2016, "What does research reproducibility mean?", *Science translational medicine*, vol. 8, no. 341, pp. 341ps12-341ps12.

Gundersen, O.E. & Kjensmo, S. 2018, "State of the Art: Reproducibility in Artificial Intelligence", *The 32nd AAAI Conference on Artificial Intelligence (AAAI 2018)*, February 2-7, New Orleans, LA, USA.

Hutson, M. 2018, "Artificial intelligence faces reproducibility crisis", *Science (American Association for the Advancement of Science)*, vol. 359, no. 6377, pp. 725-726.

Isdahl, R. & Gundersen, O.E. 2019, "Out-of-the-box Reproducibility: A Survey of Machine Learning Platforms", *15th International Conference on e-Science and Grid Computing (IEEE)*, San Diego, CA, USA,

Kondrateva, E.; Belozerova, P.; Sharaev, M.; Burnaev, E.; Bernstein, A. & Samotaeva, I. 2020, "Machine learning models reproducibility and validation for MR images recognition", *Proceedings of SPIE 11433, Twelfth International Conference on Machine Vision (ICMV 2019)*, Amsterdam, The Netherlands, pp. 114330Z.

Kratz, J. & Strasser, C. 2014, "Data publication consensus and controversies", *F1000 research*, vol. 3, pp. 94-94.

McNutt, M. 2014, "Reproducibility", *Science (American Association for the Advancement of Science)*, vol. 343, no. 6168, p. 229.

Merkel, D. 2014. Docker: lightweight linux containers for consistent development and deployment. Linux journal, 2014(239), 2.

Miller, T. 2019, "Explanation in artificial intelligence: Insights from the social sciences", *Artificial intelligence*, vol. 267, pp. 1-38.

Mitchell, M.; Wu, S.; Zaldivar, A.; Barnes, P.; Vasserman, L.; Hutchinson, B.; Spitzer, E.; Raji, I.D. & Gebru, T. 2019, "Model Cards for Model Reporting", *FAT* '19: Proceedings of the Conference on Fairness, Accountability and Transparency*, Atlanta, GA, USA, pp. 220-229.

National Science Foundation (NSF) 2020, *Proposal & Award Policies & Procedures Guide*, Available at: https://www.nsf.gov/pubs/policydocs/pappg20_1/nsf20_1.pdf (Accessed: 2020-12-29)

Nosek, B.A.; Alter, G.; Banks, G.C.; Borsboom, D.; Bowman, S.D.; Breckler, S.J.; Buck, S.; Chambers, C.D.; Chin, G.; Christensen, G.; Contestabile, M.; Dafoe, A.; Eich, E.; Freese, J.; Glennerster, R.; Goroff, D.; Green, D.P.; Hesse, B.; Humphreys, M.; Ishiyama, J.; Karlan, D.; Kraut, A.; Lupia, A.; Mabry, P.; Madon, T.; Malhotra, N.; Mayo-Wilson, E.; McNutt, M.; Miguel, E.; Paluck, E.L.; Simonsohn, U.; Soderberg, C.; Spellman, B.A.; Turitto, J.; VandenBos, G.; Vazire, S.; Wagenmakers, E.J.; Wilson, R. & Yarkoni, T. 2015, "Promoting an open research culture", *Science (American Association for the Advancement of Science)*, vol. 348, no. 6242, pp. 1422-1425.

Parsons, J. 2016, *Rogue video game AI creates 'super weapons' and leads Skynet-style uprising against human players*, The Mirror, Available at: https://www.mirror.co.uk/tech/rogue-video-game-ai-creates-8180912 (Accessed: 2020-12-23)

Peng, R.D. 2011, "Reproducible research in computational science", *Science*, vol. 334, no. 6060, pp. 1226–1227.

Plesser, H.E. 2018, "Reproducibility vs. Replicability: A Brief History of a Confused Terminology", *Frontiers in Neuroinformatics*, vol. 11, article 76, pp. 1-4.

Reese, H. 2016, *Tesla driver dies in first fatality with Autopilot: What it means for the future of driverless cars*, TechRepublic, Available at: https://www.techrepublic.com/article/tesla-driver-dies-in-first-fatality-with-autopilot-what-it-means-for-the-future-of-driverless-cars/ (Accessed: 2020-12-23)

Rocha, V. 2016, *Crime-fighting robot hits, rolls over child at Silicon Valley mall*, Los Angeles Times, Available at: https://www.latimes.com/local/lanow/la-me-ln-crimefighting-robot-hurts-child-bay-area-20160713-snap-story.html (Accessed: 2020-12-23)

Rupprecht, L.; Davis, J.C.; Arnold, C.; Gur, Y. & Bhagwat, D. 2020, "Improving reproducibility of data science pipelines through transparent provenance capture", *Proceedings of the VLDB Endowment*, vol. 13, no. 12, pp. 3354-3368.

Sandve, G.K.; Nekrutenko, A.; Taylor, J. & Hovig, E. 2013, "Ten Simple Rules for Reproducible Computational Research", *PLoS computational biology*, vol. 9, no. 10, pp. e1003285-e1003285.

Sommerville, I. 2016, *Software engineering*, 10th Global edn, Pearson, Boston

Stodden, V. 2011, Trust Your Science? Open Your Data and Code. *Amstat News*, pp. 21-22.

Stodden, V. & Miguez, S. 2014, "Best Practices for Computational Science: Software Infrastructure and Environments for Reproducible and Extensible Research", *Journal of Open Research Software*, vol. 2, no. 1, pp pp. 1-6

Stodden, V.; McNutt, M.; Bailey, D.H.; Deelman, E.; Gil, Y.; Hanson, B.; Heroux, M.A.; Ioannidis, J.P.A. & Taufer, M. 2016, "Enhancing reproducibility for computational methods", *Science (American Association for the Advancement of Science)*, vol. 354, no. 6317, pp. 1240-1241.

Tatman, R.; VanderPlas, J. & Dane, S. 2018, "A Practical Taxonomy of Reproducibility for Machine Learning Research", *2nd Workshop on Reproducibility in Machine Learning Workshop (ICML 2018)*,

Stockholm, Sweden

The Lancet Respiratory Medicine, 2018, "Opening the black box of machine learning", *The lancet respiratory medicine*, vol. 6, no. 11, pp. 801-801.

Zaharia, M.; Chen, A.; Davidson, A.; Ghodsi, A.; Hong, S.A.; Konwinski, A.; Murching, S.; Nykodym, T.; Ogilvie, P.; Parkhe, M.; Xie, F. & Zumar, C. 2018, "Accelerating the Machine Learning Lifecycle with MLflow", *Bulletin of the IEEE Computer Society Technical Committee on Data Engineering*